Geology of Construction Materials

Topics in the Earth Sciences

SERIES EDITORS

T. H. van Andel
Stanford University

Peter J. Smith
The Open University

Titles available

1. **Radioactive Waste Disposal and Geology**
   Konrad K. Krauskopf

2. **Catastrophic Episodes in Earth History**
   Claude C. Albritton, Jr

3. **Tectonics of Suspect Terranes**
   Mountain building and continental growth
   David G. Howell

4. **Geology of Construction Materials**
   John E. Prentice

Geology of Construction Materials

John E. Prentice

PHD, FGS, MInst GEOL

Emeritus Professor of Geology
University of London

CHAPMAN AND HALL

LONDON · NEW YORK · TOKYO · MELBOURNE · MADRAS

UK	Chapman and Hall, 11 New Fetter Lane, London EC4P 4EE
USA	Van Nostrand Reinhold, 115 5th Avenue, New York NY10003
JAPAN	Chapman and Hall Japan, Thomson Publishing Japan, Hirakawacho Nemoto Building, 7F, 1-7-11 Hirakawa-cho, Chiyoda-ku, Tokyo 102
AUSTRALIA	Chapman and Hall Australia, Thomas Nelson Australia, 480 La Trobe Street, PO Box 4725, Melbourne 3000
INDIA	Chapman and Hall India, R. Sheshadri, 32 Second Main Road, CIT East, Madras 600 035

First edition 1990

Typeset in 11/12 Bembo by Best-set Typesetter Ltd, Hong Kong.
Printed in Great Britain by St. Edmundsbury Press, Bury St. Edmunds, Suffolk.

ISBN 0 412 29740 X
0 442 31162 1 (USA)

British Library Cataloguing in Publication Data
Prentice, J. E.
Geology of construction materials.
1. Building materials. Geological aspects
I. Title II. Series
691'.2

ISBN 0 412 29740 X

Library of Congress Cataloging in Publication Data available

Contents

Series foreword

Year by year the Earth sciences grow more diverse, with an inevitable increase in the degree to which rampant specialization isolates the practitioners of an ever larger number of subfields. An increasing emphasis on sophisticated mathematics, physics and chemistry as well as the use of advanced technology have set up barriers often impenetrable to the uninitiated. Ironically, the potential value of many specialities for other, often non-contiguous ones has also increased. What is at the present time quiet, unseen work in a remote corner of our discipline, may tomorrow enhance, even revitalize some entirely different area.

The rising flood of research reports has drastically cut the time we have available for free reading. The enormous proliferation of journals expressly aimed at small, select audiences has raised the threshold of access to a large part of the literature so much that many of us are unable to cross it.

This, most would agree, is not only unfortunate but downright dangerous, limiting by sheer bulk of paper or difficulty of comprehension, the flow of information across the Earth sciences because, after all it is just one earth that we all study, and cross fertilization is the key to progress. If one knows where to obtain much needed data or inspiration, no effort is too great. It is when we remain unaware of its existence (perhaps even in the office next door) that stagnation soon sets in.

This series attempts to balance, at least to some degree, the growing deficit in the exchange of knowledge. The concise, modestly demanding books, thorough but easily read and referenced only to a level that permits more advanced pursuit will, we hope, introduce many of us to the varied interests and insights in the Earth of many others.

The series, of which this book forms a part, does not have a strict plan. The emergence and identification of timely subjects and the availability of thoughtful authors, guide more than design the list and order of topics. May they over the years break a path for us to new or little-known territories in the Earth sciences without doubting our intelligence, insulting our erudition or demanding excessive effort.

Tjeerd H. van Andel and Peter J. Smith
Series Editors

Preface

Almost all building materials used by man – if we except timber – are based upon minerals. It is therefore very surprising that geologists, whose science is most concerned with mineral matters, have in the past interested themselves little in the basic raw materials of the construction industry. It has been left to chemists, for instance, to unravel the mineralogy of cement; and it has been engineers, both civil and mechanical, who have developed the schemes for testing and standardization of these mineral products. The author is happy to say that, over the past few decades, that situation has changed. The increasingly competitive search for new mineral sources has led the extractive industry to employ more and more geologists, and so their involvement with all aspects of the mineral industry, from extraction to utilization, has increased.

The chief contribution of the geologist to this field is the recognition that every mineral or rock owes its properties to its fundamental origin. There is thus a direct connection between the processes of mineral formation in the earth's crust and the mode of occurrence and essential properties of the mineral. Although the connection between the imaginative and thought provoking theories of plate tectonics and the humdrum use of, say, sand in a mortar mix, is tenuous, it is nevertheless real. The geologist needs to use this fundamental understanding to bring a rationality to the practical purposes.

In this book I have attempted to show that the geological approach is worth while and rewarding, as giving a coherence and logic to an otherwise chaotic and undisciplined subject. I have assumed a basic understanding of geological principles, while avoiding, as far as possible, abstruse jargon – to which geologists, like many other scientists, are much addicted.

The book is intended to be useful to undergraduates in geology and related disciplines, who wish to know more of a subject which is unlikely to figure prominently in their regular courses; it is hoped it might interest them sufficiently to seek a career in the field, or even to undertake research projects in the many areas which need investigation. It should also be of use to young graduates who find themselves, probably unexpectedly, working in the extractive industry. I would hope that it will also find its way into the hands of managers in the industry, and perhaps show them the ways in which the expertise of their geologists could be used and extended.

It is not intended that this book should be a working handbook or reference source for those actively engaged in the industry; many subjects discussed herein are much more complex than this inevitably superficial account suggests. This need has been filled for the aggregates industry by the *Geological Society Engineering Geology Special Publication No. 1 – Aggregates*, an invaluable work which has as yet no counterparts in other areas of the extractive industry.

In a book of this size and character, proper acknowledgement of the basis of each and every statement is impossible; apologies are thus due to the authors of the many research publications on which I have so extensively depended, who may well recognize their own work, but for which I have been unable to give appropriate credit.

In attempting to cover such a wide and varied field as this, the author is only too well aware of the extent of its deficiencies. Some of these deficiencies may be due to the wide range of publications in which research in the field is published – there may well be many vital pieces of information published in the many scientific, technical and trade journals which the author has missed. In some cases the information is not available – complete inventories of many industries are lacking in many parts of the world, for instance.

Inevitably, much of the matter and the examples used come from the United Kingdom; in many cases this is typical of the situation in western Europe and north America – where there are important differences these are noted. The developing countries are different again, and I have attempted to note these differences where possible.

It is my greatest hope that this book will, by showing the connections between 'academic' geology and the needs of the extractive industry, do something to improve communication at all levels. Demands upon the construction industry grow greater all the time; at the same time raw materials become scarcer. The proper use of our mineral resources can only be achieved if we use all our geological skills to ensure that they are used to the best advantage.

I owe a great debt to the many persons, both in industry and in the academic world, who have contributed to my education in this large and complex field of study. I must thank in particular many friends and colleagues in Redland Ltd, on whose projects I have gained much of my industrial education, and one who starts as a specialist geologist must learn a great deal, and this learning comes from everyone from the Managing Director to the drag-line operator. Above all, however, my thanks go to my wife, Pat, without whose forbearence, encouragement, and continued help this book would never have been started, persevered with, or completed.

1 Introduction

1.1 CONSTRUCTION MATERIALS PAST AND PRESENT

Man has always used natural rock and stone for his own purposes – indeed, it was his ability to do so which distinguished early man from his ape-like ancestors. At first it was only for weapons and tools; but early in his history man began to use rock for building; at first to improve his cave or shelter, but later to construct his dwellings. Although it must be assumed that timber was probably the commonest building material used by early man, building in stone goes back at least to neolithic times. At Tell es-Sultan, near Jericho, undressed stone was built into town and house walls around 6000 BC. In the dry climates of the Middle East and south-east Europe, many walls were of earth-plaster, and at Catal Hüyük, in Turkey, built-in furniture was also constructed (at about 6800 BC) of the same material. Mud-bricks, moulded from clay but unburnt, were also made at this early date. In northern regions like those most recently deglaciated, where timber was scarce, and mud walls could not survive, stone construction reached a high degree of sophistication, as for example at Skara Brae, in Orkney (Figure 1.1), where, around 2000 BC, the neolithic people constructed walls, benches, beds and cupboards from the local flagstones; and the angularity and lack of weathering shows that the slabs were deliberately quarried.

The Mediterranean civilisations of Egypt, Greece and Rome were responsible for great advances in the use of natural stone for building. Whereas the Mycenaean civilisation (*c.*1500 BC) constructed 'cyclopean' walls (Figure 1.2), in which unmodified blocks were fitted neatly together; the Egyptians had earlier (around 3000 BC) than this learnt to shape their comparatively soft Tertiary limestones into rectangular blocks, and by 2000 BC had been able to apply this process to granite. Greece and Rome, of course, developed masonry building in shaped blocks to a fine art. Since that time, 'dimension stone' has occupied a primary place in building construction, especially where buildings of size and prestige are concerned.

Figure 1.1 Skara Brae, Orkney; interior of neolithic house.

In none of these early buildings, where shaped stones were used, does mortar play a significant part; it was used, but apparently largely to slide blocks into place rather than as a cement. It played a more important role where unshaped stones were used, and rendering of such walls was commonplace from earliest times. In Egypt, gypsum was the major constituent of such renderings, while the Romans and Greeks used burnt lime and clay. These people also discovered the cementitious properties of volcanic ash, and all these various minerals, alone or in combination, continued to be used until the discovery of Portland cement in 1824. The high strength of this latter material made the production of concrete possible. Made by the incorporation of stone particles – aggregate – both coarse and fine in a cement matrix, concrete plays today a major and indeed essential role in the construction of buildings large and small, and in roads and bridges. The demand for cement, and for aggregates for road and building construction, is filled by naturally occurring gravels, and by crushed rock, and demand continues to rise throughout the world.

Clay, shaped and burnt into bricks, also has a long history, some

Figure 1.2 'Cyclopean' masonry, Ancient Asine, Greece (photograph: JEP).

examples being known from around 1200 BC, although the Romans must be credited with the real development of this product. Throughout history, and in almost all parts of the world, clay bricks and tiles have been a most important material for the building of walls, roofs and floors in houses large and small; and there are no indications that their use is likely to diminish in the foreseeable future.

The construction of roads from unbound rock fragments has been a usual technique since Roman times at least, although in medieval and later times it seems largely to have been forgotten, and it is to James Macadam (1756–1836) that the beginnings of modern road construction must be ascribed. Today, roads are built with a sub-base of mainly coarse aggregate, and surfaced by concrete or asphalt. Originally a naturally occurring material in Trinidad, Cuba and Switzerland, asphalt is now made from incorporating fine aggregate in a bituminous mix. The surface of the road is coated by further aggregate – usually crushed rock. Thus a very large proportion of the road consists of aggregates of one sort or another – and the demand created by the construction of a modern highway system is clearly enormous.

Our modern society, in both the developed and the developing world, continues to ask for higher standards of housing, better roads, more and larger bridges, more airports. The mineral extractive industries must strive to supply the materials for these developments. Attempts to substitute other materials – metals such as aluminium, plastics and by-products made from waste – for the traditional mineral-based products, make little inroads, since these alternative materials are always either too costly, or less satisfactory in use. There is no escape from the conclusion that the construction industry must continue to depend on rocks and minerals for its basic needs.

However, while we must recognize the need for the supply of rocks and minerals, we are not always tolerant of the disturbance to the environment that this causes. Most of the industrial minerals and rocks required by the construction industry are won by open-pit extraction; deep-mining methods are in general too expensive, and are only resorted to when the mineral is particularly rare, or where it occurs only in thin veins. As an instance of the first case are certain Italian marbles of specific colours and types, which are followed from the surface by mining methods; an example of the second are some gypsum deposits in south-east England, which occur too deeply beneath overburden to be worked by opencast methods.

Open-pit extraction, however, uses a great deal of land surface. This is particularly true of many sand and gravel deposits, producing coarse and fine aggregates, which often occur as river terraces of only a few metres in thickness – thus their extraction rapidly removes large areas of land. Hard rock for crushed stone, or construction stone, and brickclays tend to be won from deeper excavations, thus having less effect upon the surface. Quarrying operations, however carefully conducted, are often environmentally unacceptable. Noise from blasting, from heavy machinery, and from rock crushing; dust from crushing and screening; and the passage of heavy vehicles in and out of the quarry – all these are difficult to accommodate if people live and work nearby.

Moreover, the effects of open-pit extraction may last long after mineral extraction has ceased. In areas of river gravel, for instance, where the water-table is very near the surface, there is little alternative to leaving the excavation as open water. Where the excavation is mainly dry, an acceptable solution might be to fill the quarry. However, material to fill a quarry is not always readily available; often the only available material is domestic waste. Domestic waste filling can itself be an offensive process; and if the base of the quarry is not effectively sealed from any groundwater, contamination of surface water, and even of water-supply aquifers, can occur.

Thus there is a fundamental contradiction. Society demands the provision of high-volume, low-cost minerals and rocks to build its houses and roads; at the same time it is not prepared to accept the loss of land and amenity, nor to tolerate the environmental consequences of the quarrying process. The consequences of this contradiction have been appreciated by most developed countries, who now have elaborate systems of planned control of mineral working, designed to strike an acceptable balance between these conflicting demands. Rarely is such control exerted in less developed parts of the world. In the future, there is no doubt that we shall see an extension of restriction of areas in which minerals can be worked, which will in turn lead to a search for these bulk minerals in areas further from their place of use. That this will add substantially to the cost of construction is manifest; but this is the price which will have to be paid if the environmental consequences of quarrying are to be contained.

All the minerals and rocks used by the construction industry are abundant and of common occurrence, and there is certainly no suggestion that there is, or is likely to be, a danger that they will be exhausted. There is, however, a clear possibility that there will be local shortages, and because the cost of transport is so large a factor, these shortages may be difficult to remedy by bringing in material from further away. A good example is the situation with regard to concrete and road aggregates in the south-east of England. This region consumes the largest volume of these materials in the British Isles. Traditionally it has relied upon the terraces and alluvial plain of the River Thames to supply these aggregates; but this source is subject to intense pressure. Water-table in them is high, so the extensive workings of the past century are mostly left as areas of water, and while some of these are used for recreation, for bird sanctuaries and the like, there is much resistance to their extension. Some large areas of reserve exist, but they have been sterilized by the unplanned extension of housing across their surface. And other demands for the land, for housing, for motorways, and for airport extensions, are strong. Thus the region expects that this resource will disappear before the end of the century, and the region will be obliged to seek its aggregate from sea-dredged materials, and from crushed rock imported from as far away as Scotland, with the inevitable increase in cost that such long hauls will bring. A similar situation is developing in the north-east United States; there is no shortage of hard rock, but it occurs in areas which are being increasingly populated, or which are relied upon for the large urban population for recreation. Thus the development of coastal quarries situated many miles away, but from which aggregate can be brought by sea, is an inevitable consequence.

It follows from this that there is a need to ensure that supplies of raw material for the construction industry are used wisely, and in a manner most advantageous to the community. The geologist's contribution to this is to provide a fundamental understanding of these materials; his or her responsibilities are:

1. to ensure that the minerals and rocks supplied to the construction industry are suitable for the purpose for which they are required, and
2. to ensure that a continuing supply of such materials is made available to the industry.

To satisfy condition 1 requires an understanding of the properties of the raw material, and a knowledge of how that material will behave in use. For instance, it may be necessary to know that a particular aggregate to be used for, say, a concrete road-surface, will have the qualities of strength, durability and ability to resist polish, for the type of road for which it will be used, and that it will neither react adversely with the concrete, nor disintegrate under the weathering conditions of the particular region. Or it may be needed to know that a certain clay will make the kind of brick that is currently saleable in the region; or that a particular gypsum can be made into a plaster wallboard. Since all these properties depend, in the last analysis, upon the precise mineralogy of the stone, the clay, or the gypsum, they are very properly within the remit of the geologist.

Satisfying condition 2 needs a thorough understanding of the rock or mineral deposit; its shape in the ground, the quantity of mineral it contains, the extent of its internal variability, are all consequences of its method of formation. The geologist's training is very largely directed to understanding the processes by which such deposits form, and so he or she is uniquely qualified to assess these factors, and hence to determine the contribution which a particular deposit can make to the supply of the particular raw material.

The work of the geologist in the provision of minerals and rocks for the construction industry falls into the following categories:

1. Prospection – the location of new areas where the mineral is likely to occur, and which are worthy of further investigation;
2. Exploration – the testing of the deposit, usually by some means of subsurface investigation such as boreholes, trenches etc.; and the collection of samples for laboratory study;
3. Assessment and evaluation – the calculation of quantities, or volume of mineral; the assessment of quality from the laboratory tests; and the appraisal of the economic viability of the project;
4. Exploitation – the recommendation of the working method for the quarry; the planning and design of the quarry, and the continued safety of the operation; the geologist should also advise on any environmental consequences of the operation – e.g. possible pollution or interception of groundwater etc;
5. Restoration – the planning of an appropriate scheme of restoration for after-use of the quarry, and the probable environmental effects thereof.

At all times during this long, and often quite expensive, process the geologist needs to keep economic considerations firmly to the forefront, and he or she needs to build into their plan a series of points at which decision to proceed or not is made, so that expenditure does not proceed beyond points where economic viability has ceased to be possible. Above all, the geologist must remember that everything depends upon the correctness of the geological interpretation. It is never possible to prove, with 100% certainty, either the quantity or quality of a deposit; there is always the possibility that some geological situation has been misinterpreted – the unexpected thick lens of clay in an alluvial deposit, the zone of shattered rock which all the boreholes missed, are examples. So geologists need to be constantly aware of this, and to be regularly asking themselves the question – 'what happens if I am wrong?'.

At the same time, the geologist should never be afraid to ask for more expenditure on exploration if, in his or her professional judgement, it is necessary. In an industry which produces a low-cost material, this is rarely popular; but if geologists are convinced that failure to spend more on exploration could fundamentally change the economics of an operation, they should not hesitate to say so. They are constantly being asked to give their judgement on a balance of probabilities; this judgement can only be based on their training and experience; but this is of no value if the basic information is incomplete.

1.2 FINDING, EXPLORING AND ASSESSING

The responsibility for bringing a new discovery or prospect to the point where it is a profitable quarry must rest with the geologist. The following section outlines some of the procedures which a geologist needs to follow in seeking for, planning and developing such a prospect.

The earliest stage is the location of areas which are likely to contain the rock or mineral – the prospecting stage. In the field of construction minerals, the overriding factor is always distance from the market. Minerals for the construction industry sell at very low prices from the pit gate, so that transport costs form a very large part of the final selling price. It is said that the cost of sand and gravel doubles for every ten miles of road-distance from source to market. Normally the market is a fixed point, so that the prospecting area is delimited by a fixed radius from that point. Very occasionally, the geologist may be seeking a supply source more widely, so that a factory using this may be established close to the source – but this is exceptional. Longer distances can be tolerated if some form of water-transport – sea, river or canal – is adjacent to both source and market. In general, however, the limits of possible exploration are closely defined by this factor.

Choice of area for prospecting is often limited by other non-geological factors; by the presence or otherwise of a metalled road, by availability of fuel, and of water supply. In developed countries, the probability of being able to lease or purchase the land; and the likelihood of obtaining planning permission or a mining licence all have to be taken into account. The net result is that the geologist is likely to be seeking the mineral in a fairly closely limited area.

An important step at this stage is to define precisely the material that is sought. If it is a construction stone, what kind of strength, hardness and appearance are being sought? If an aggregate, what kind of granulometry or mineralogy? Are there any constituents which must be avoided? The importance of this step is that it enables the geologist to define the depositional environments in which such a deposit might be found, and thus further narrow the search.

When he or she has established the limits of the prospecting area, and defined the precise objectives of the search, the geologist's next action will be to consult the local geological map. The availability of detailed geological mapping is enormously variable. Even within a tolerably well-mapped country such as Britain, there can be remarkable anomalies. Thus a geologist seeking, say, sand and gravel supplies in eastern England, will find that parts of the area are covered by detailed 1:25 000 maps, with all outcrops recorded, and with detailed borehole records fully tabulated and analysed; while adjacent areas have no map more recent than the first 1 inch to one mile survey in 1890, which does not always even record the presence of drift deposits. It is rare to find no geological map at all, although there are parts of the world where this is so; and there are large areas where the geological map is only of a very large scale. Almost without exception, however, all countries have a Geological Survey organization of some kind, and their office, or their publications, should be the first to be sought.

At the same time, there is a vast amount of useful information contained in the voluminous academic literature on geology, in the pages of the many scientific and technical journals. A working knowledge of the major publications, and the skill to conduct a 'library search' is thus invaluable. Computer searches of the available databases in geology are now a useful starting point, but they do not carry data before 1967 (or 1974) – and often vital information is contained in publications back into the last century. This information is not always related to mineral extraction, but always provides useful background which makes the geological judgement that bit more sound.

An illustration of the above process was a search carried out to locate a supply of quartzose sand for a proposed concrete products factory in Greece. The remit was to find such a supply within a reasonable distance of a major centre of population; the sand had to conform to a closely specified granulometry, and to be free from particles of chert or basic igneous rocks; not more than 30% of the 'sand' could be of limestone particles. Study of maps and academic literature on Greece revealed that most of the country was a highly tectonized sequence of nappes, in which limestone, ophiolites and cherts were ubiquitous. Any chance of obtaining the sand from the tectonized rocks was ruled out, since all the sandstones in this sequence were essentially greywackes, and all strongly lithified. Outside the tectonic mountains, there are plains consisting of Neogene sediments; but bearing in mind the climatic history of the region in the Neogene, where chemical weathering will have prevailed, these seemed likely to be largely clay. The only hopeful prospects were the alluvial plains of rivers; but study of the map showed that most of the major rivers drained from the tectonic terrain, and the likelihood that their sands were free from limestone, chert and ultramafic igneous particles was remote. This narrowed the search to a group of rivers draining southwards from the granitic Rhodope massif in northern Macedonia, whose alluvial sands seemed likely to contain more quartz. All these prognostications were made before leaving England; a rapid reconnaissance in southern Greece soon confirmed their accuracy, and an appropriate sand was found in the alluvium of the Gallikos River near Thessalonika.

When the prospection phase has been completed, and thus when an area or areas which are deemed worthy of further investigation have been located, the next stage is to carry out an exploration. While surface indications, which form the basis of most geological mapping, will give a first indication of the presence or absence of a particular mineral, any proper assessment of quantities must rest on subsurface exploration.

If the deposit is shallow, excavation of trial pits by mechanical excavator is satisfactory; the method is cheap and quick, and provides the geologist with an opportunity to see the deposit in three dimensions; it also provides the opportunity to collect large samples. Such trial pits are limited to the reach of the excavator arm, and can thus rarely be extended beyond 3 m from the surface; deeper excavation requires permanent shoring of the shaft, and is extremely expensive – rarely justified in the case of a bulk mineral. Deeper exploration therefore requires the making of boreholes.

The cheapest method of borehole construction is **percussive cable boring** (oddly, and misleadingly, called 'shell and auger') (Figure 1.3). The method only requires a simple power winch, which is used to hoist and drop a heavy hammer, which forces a steel tube into the ground. The same winch then raises and drops a heavy chisel, which pulverizes the material inside the tube – this is then drawn out by a bucket with a hinged base. This is called a disturbed sample – a so-called undisturbed sample can be obtained by hammering a further tube down inside the first, and then drawing this to the surface. The method is messy, the samples are mixed and often contaminated, and the large volumes of water used can produce anomalous results (see below). Nevertheless its low cost ensures its continuance in Britain, although it is rarely seen outside.

All other methods of drilling are rotary (Figure 1.4) – that is, they depend upon a rotating drill-head, which grips a string of hollow drill-rods. Rotation speeds can range from very low to up to around 600 revolutions per minute, depending upon the method; and a load of several tonnes can be applied to the drill string.

For unconsolidated deposits, a more satisfactory method than that of percussive boring is the **continuous-flight auger** (Figure 1.5). In this method, lengths of auger – generally about 150 to 200 mm in diameter – are screwed into the ground for a distance of a metre or

Figure 1.3 Percussive cable rig (photograph supplied by Laud and Marine Engineering Ltd.).

Figure 1.4 Rotary drilling rig (photograph supplied by Diamant Boart Craelius Ltd).

so, then withdrawn; the sample is retained in the blades of the auger. There is some danger of contamination by material falling from, and scraped from, the higher parts of the hole; nevertheless with care and experience good samples can be obtained. It remains one of the few methods by which reasonable sampling of unconsolidated sands and gravels can be effected.

A modification of the above uses a **hollow-stem auger** (Figure 1.6), in which the centre of the auger is a hollow pipe, often provided with a cutting edge at its base. This is forced into the ground by the downward pull on the auger vanes, and in weak rocks can provide a relatively undisturbed sample.

In harder rock, and where less disturbed samples are required, recourse may need to be had to **rotary core-drilling**. Here, a cylin-

Figure 1.5 Continuous-flight auger.

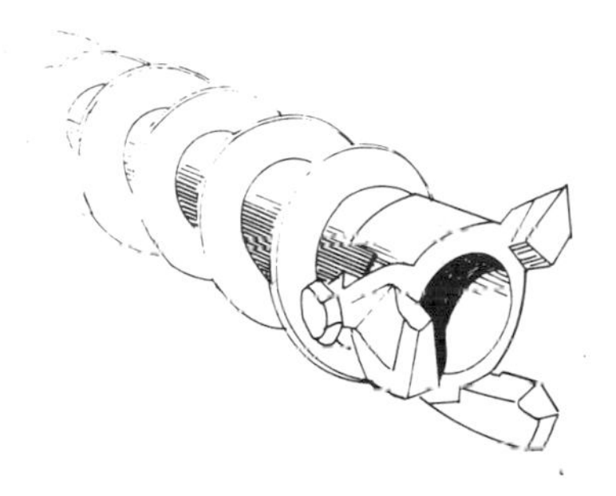

Figure 1.6 Hollow-stem auger.

drical bit is used to cut a cylindrical core from the solid rock, which is retained in a core barrel within the drill pipe. At regular intervals the whole drill string is drawn to the surface, and the core removed from the barrel. The bit needs to be armoured to cut through abrasive rock; this armour can be hardened steel (for clays and soft rocks), tungsten carbide, natural or synthetic diamond for harder rocks. The method is slow, laborious, and expensive – especially since bits need frequent replacement – and although in theory it should produce the best samples, this is not always the case. The necessity to cool the bit, and to remove the products of the abrasion, demands that the hole must be continuously flushed. The flushing medium can be compressed air, water, or a dense liquid; but its use can often erode and remove substantial parts of the core. Thus in a sequence such as may be found in many brick-clay projects, where clays alternate with poorly consolidated sands, the clay may provide a good core, but the sands are completely washed away. Moreover, if the rock being drilled is fractured, the core may break up in the core barrel, and in its rotation destroy large parts of the core; in these circumstances the core often slips from the core barrel while being drawn to the surface. While modern drilling technology can do much to reduce core loss, it still remains true that 100% core recovery is rarely attained.

In these circumstances geologists often turn to the less expensive **open-hole rotary drilling** (Figure 1.7). This uses a bit which breaks

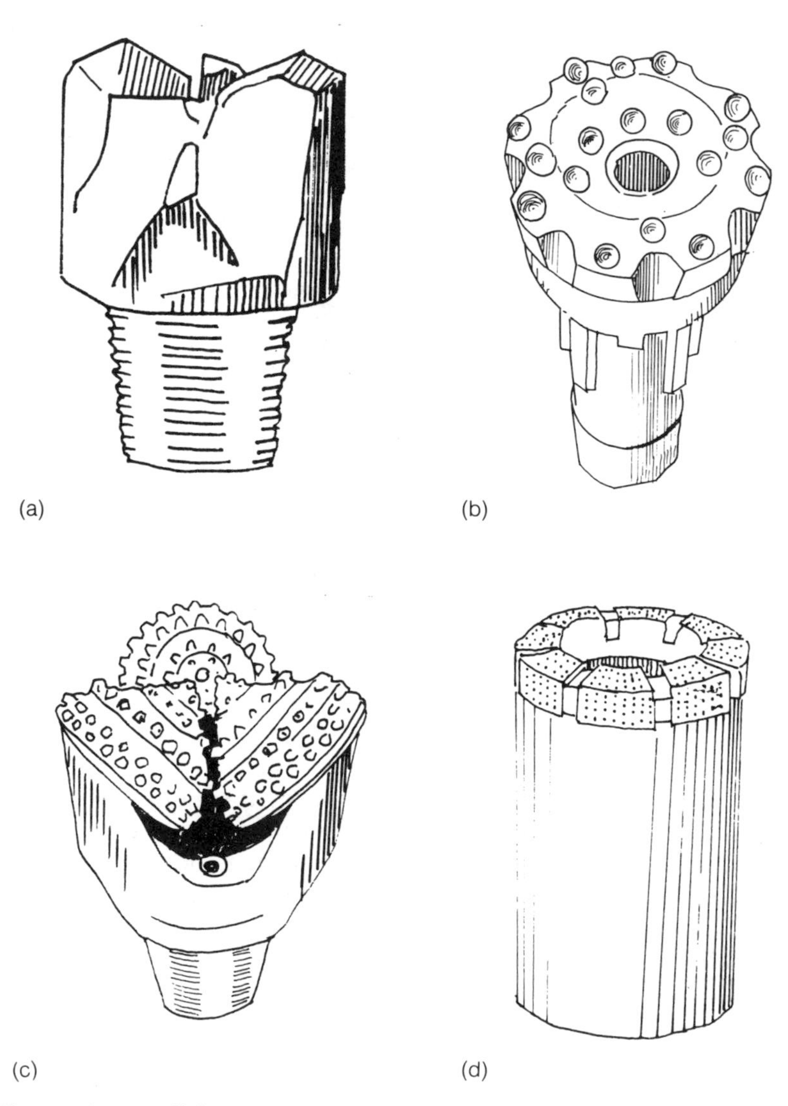

Figure 1.7 Drill-bits used in rotary drilling: (a) drag-bit; (b) button-bit; (c) tricone-bit; (d) diamond core-bit.

up the rock in its downward progress, so that samples are brought to the surface as a stream of chippings in the drilling fluid. The bits may be provided with blades of various configurations, or with angled wheels (rock-rollers), or discs; and they may be armoured with tungsten carbide or diamond inserts. Changes in lithology are identified by examination of the chippings, and thus a stratigraphic section can be built up; the samples may be adequate for laboratory analysis, but not for any form of product testing.

Open-hole drilling can, however, be used to provide further information. The use of down-the-hole television cameras to examine the sides of the borehole has proved, for instance, the only satisfactory method to explore the Sherwood conglomerate – this is a Triassic formation in the English Midlands, which consists of very hard pebbles in a weak sandstone matrix. It is too hard to auger, and the contrast of pebbles and matrix makes it impossible to core.

Careful observation, and instrumentation, of the rate of penetration of the bit can give accurate identification of lithological changes during drilling. After the drill is withdrawn, geophysical probes can be lowered down the hole, and their records used to supplement the chipping information. Particularly useful methods in the field of construction raw materials have been those which observe natural gamma-radiation, resistivity and density. Gamma-radiation in rocks emanates from radioactive minerals, which are usually present in clays, but absent in sands. Thus a gamma-log identifies precisely the extent of individual clay and sand seams – a method which has been found particularly useful in the Etruria marl (Carboniferous) of central England – a formation where the irregular distribution of thin sand layers in the sequence is a major hazard in the brickmaking process. Resistivity measurements are linked to porosity, and have been used to detect the base of sand and gravel deposits, and the position of clay-rich seams within them. Thus open-hole drilling, with careful instrumentation and logging, can be used to build up a very complete geological picture. Compared with core drilling, it is much faster and much cheaper; and it is likely to be the method increasingly favoured in the exploration of bulk mineral deposits.

Perhaps the greatest problems are encountered in the exploration of loosely consolidated deposits, such as sand and gravel. Any form of core recovery is out of the question, so the geologist must fall back on direct excavation of trial pits, on the so-called shell-and-auger method, or on some kind of augering technique

No method is ideal, and the geologist needs to execise much discretion in the interpretation of the results. A good example occurred in the investigation of a potential asphalt-sand in Northamptonshire.

The deposit, which is of fluvioglacial origin, was first explored by trial pits, in which it was observed that the sand itself was free from clay, but within it were thin seams and lenses of a silty clay material. In taking the samples for analysis, these clay bands were rejected. The deposit was then tested by shell-and-auger boreholes, and the analytical results revealed a wide difference from those of the trial pits. In consequence another set of boreholes was made, using a

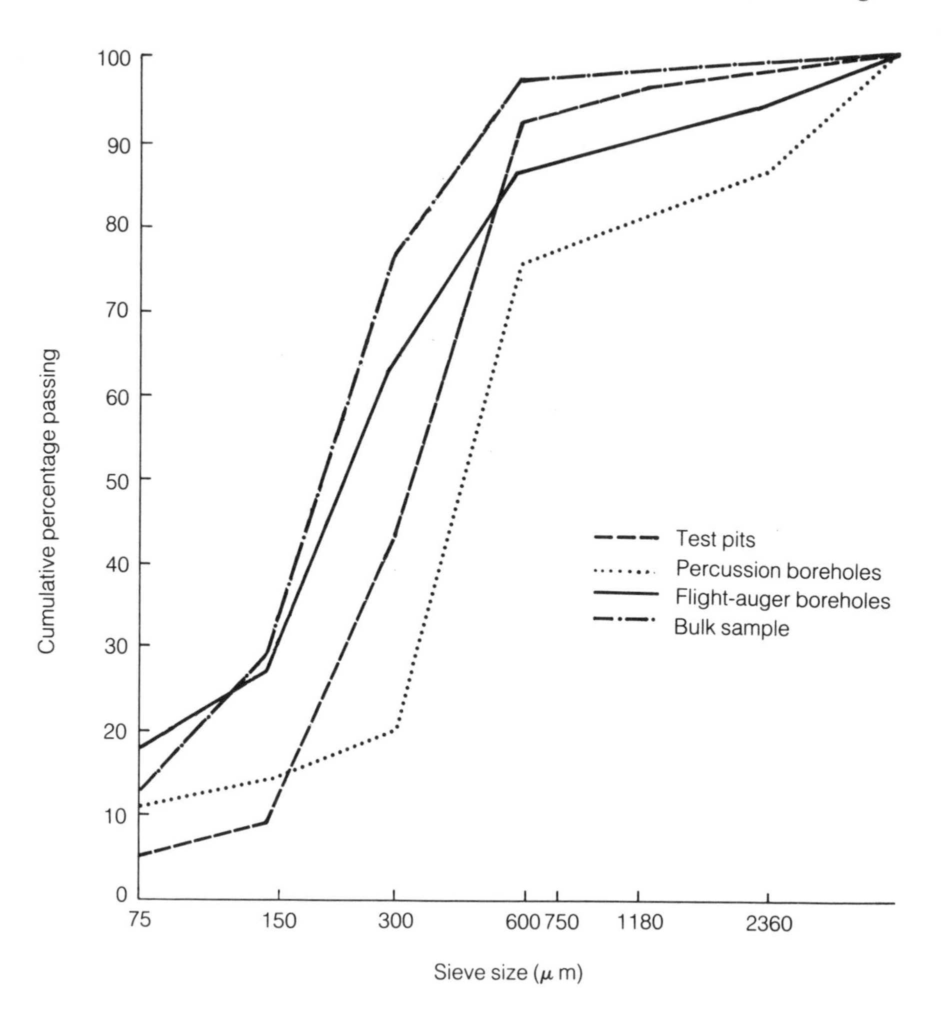

Figure 1.8 Comparison of test results: cumulative curve of sand analysis. Fluvioglacial sand at Stoke Albany, Northamptonshire. ------, excavator samples; . . . . . ., shell-and-auger samples; ———, continuous-flight auger samples; —.—.—, bulk sample.

continuous-flight auger, producing a third set of analytical results – all three are shown in Figure 1.8. Finally, a 100-ton bulk sample was collected from an excavation.

The explanation of the inconsistent analytical results lies in the methods of sample retrieval. In the trial-pit samples, the clay seams were rejected, whereas in the continuous flight auger boreholes, these seams were ground up and incorporated fully into the sample. In the shell-and-auger holes, the fine sand in the mid-range was washed out in the baling process.

The geologist has to decide which set of results to use, and he or she has to decide which method of sampling gives results which will match most closely the final, commercial product of the quarry. In this respect the excavator samples match the product of a dry-screening process, which would reject the lumps of clay: the shell-and-auger samples resemble the product of a wet-dredged sand: and those from the flight auger simulate the conditions in which the sand is put through a washing plant. The differences are not insignificant, since only the excavator sample results lie within the British Standard Specification for asphalt wearing courses.

The evaluation of bulk mineral deposits is in terms of quantity and quality: in all cases it is the spatial distribution, in three dimensions, which is important, so the construction of geological maps and sections plays a primary role in evaluation.

As a means of understanding, and then of quantifying, a mineral deposit, the contoured map is of paramount importance. The contours may be surface contours, subsurface contours on a particular geological horizon, or contours of equal depth of deposit (isopachs).

A simple example is illustrated in Figure 1.9, which records the results of an investigation, using continuous-flight augering methods, into an area of alluvially deposited sand. Eighteen boreholes were made, three of which stopped short of the base of the deposit. Since the surface is effectively flat, it is adequate to use surface level as a datum in this case, so that simple 'depth to base of sand' can be used to construct the contours. The depth to base is shown against each borehole position in Figure 1.9(a); 5 m, 10 m and 15 m points are found by intercalation. It is important to appreciate that the validity of this method is dependent on the assumption that there is an even and continuous slope between adjacent boreholes.

The intercalated points are then joined to form contours; the pattern would be a simple one if it were not for three boreholes in the centre, which show anomalously shallow depths of 9.0, 5.7, and 3.0 m. One interpretation is shown in Figure 1.9(b). This explains the apparently anomalous shallow depths in the centre of the map by

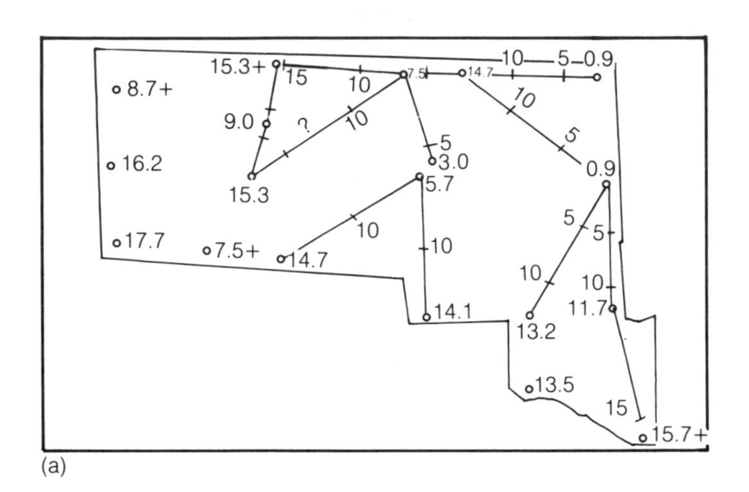
15.3+
15
10
7.5
14.7
10
5
0.9
8.7+
9.0
?
10
10
5
5
3.0
16.2
15.3
5.7
0.9
10
5
5
17.7
7.5+
14.7
10
10
10
14.1
11.7
13.2
13.5
15
15.7+
(a)

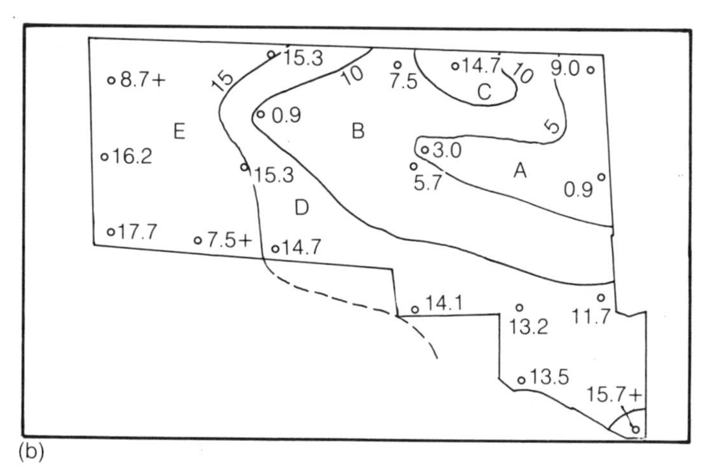
15.3
15
10
7.5
14.7
10
9.0
8.7+
C
0.9
E
B
5
16.2
3.0
A
15.3
5.7
0.9
D
17.7
7.5+
14.7
14.1
13.2
11.7
13.5
15.7+
(b)

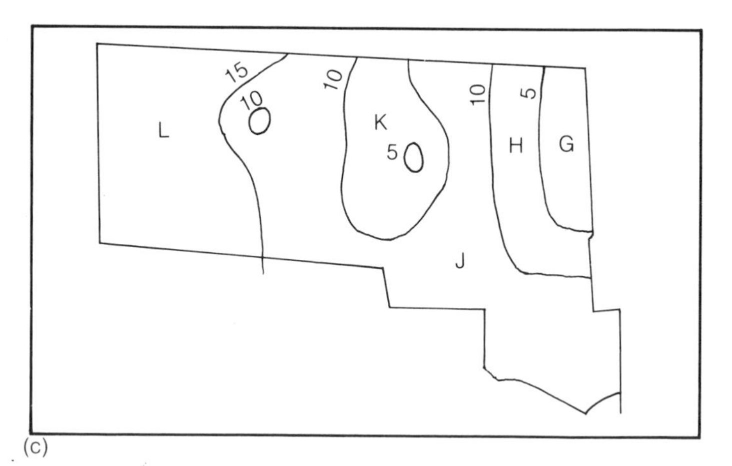
15
10
10
10
5
L
K
5
H
G
J
(c)

postulating a central east–west ridge. To do this, it has to be assumed that the intercalation of a 10-m height (ringed in Figure 1.9(a)) between the boreholes with depths 7.5 and 15.3 is invalid. An alternative explanation, which uses all the intercalated points, but which postulates two curious humps around the three anomalous boreholes, is shown in Figure 1.9(c).

The procedure by which these contours can be made to provide volumes or quantities involves the measurement of the area between contours, which is best carried out with a planimeter; this device, which was originally mechanical, but which can now be linked to a computer, measures the area of irregular shapes, and is usually provided with adjustments which allow direct conversion of the scale of the map to the area on the ground. If these areas are then multiplied by the appropriate mean depth (e.g. 12.5 m for the area between 10 and 15 m contours), a good approximation of the volume

Figure 1.9 Calculation of quantities of sand in an alluvial deposit (scale 1:2500): (a) location of boreholes, with depths to base of sand (in metres), and intercalated points for contour construction: ○ 14.7, borehole with depth to base of sand in metres; ○ 8.7+, borehole depth, base of sand not reached; (b) as above, contours constructed on first interpretation; (c) as above, alternative interpretation.

a *Area no.*	*b* *Planimeter reading*	*c* *Multiply by (scale factor)*	*d* *Area* (m^2) *(b × c)*	*e* *Depth (m)*	*f* *Volume* (m^3) *(d × e)*
A	0400	50	20 000	2.5	50 000
B	1089	50	54 450	7.5	408 375
C	0121	50	6 050	12.5	75 625
D	1123	50	56 150	12.5	701 875
E	0945	50	47 250	16.0	756 000
Total					1 991 875
Alternative interpretation					
G	0196	50	9 800	2.5	245 000
H	0365	50	18 250	7.5	136 875
J	1692	50	84 600	12.5	1 057 500
K	0421	50	21 050	7.5	157 875
L	0996	50	49 800	16.0	796 800
Total					2 173 550

NGR: BB 9744.
Location: Ambridge Hall Farm, Ambridge, Loamshire (north-east fields 180, 187, 193).

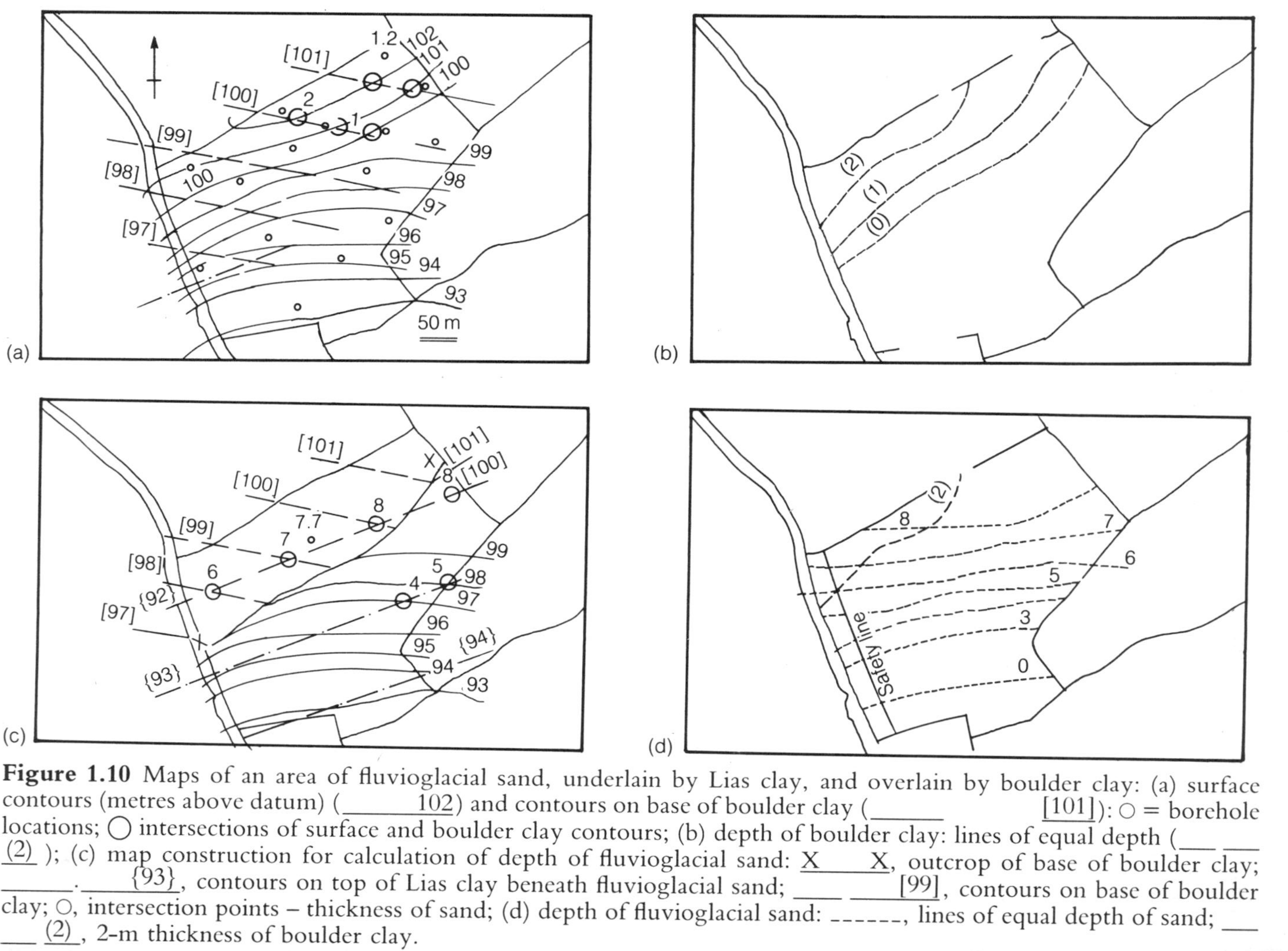

Figure 1.10 Maps of an area of fluvioglacial sand, underlain by Lias clay, and overlain by boulder clay: (a) surface contours (metres above datum) (_____ 102) and contours on base of boulder clay (_____ [101]): ○ = borehole locations; ◯ intersections of surface and boulder clay contours; (b) depth of boulder clay: lines of equal depth (___ ___ (2)); (c) map construction for calculation of depth of fluvioglacial sand: X____X, outcrop of base of boulder clay; _____._____ {93}, contours on top of Lias clay beneath fluvioglacial sand; ____ ____ [99], contours on base of boulder clay; ○, intersection points – thickness of sand; (d) depth of fluvioglacial sand: ------, lines of equal depth of sand; ___ ___ (2), 2-m thickness of boulder clay.

can be obtained. Calculations based on each of the two interpretations are given in Figure 1.9; and it will be seen that there is a difference of some 180 000 m^3 in the results.

At this stage the geologist needs to decide whether the possible error involved justifies calling for more boreholes; and he or she is required to give an opinion on

1. which of the two interpretations is correct, and
2. whether there are other interpretations which might be equally valid.

Now this particular sand is known to have been deposited by a substantial river flowing from east to west, and the base of the channel is known to be cut in bedrock. Given this background information, the geologist can say with some confidence that the first interpretation is more likely to be correct than the second. Other possibilities, such as that there exist a series of small humps in the base, only two of which have been encountered fortuitously in boreholes, which might be a possibility if the bedrock were moraine, can reasonably be dismissed; but all need to be considered. Unfortunately, in geology, the simplest interpretation is not always the correct one.

Assessment becomes more complex as the number of variables increases. This occurs

1. when the ground-surface is not flat
2. when both upper and lower surfaces of the deposit are variable, and
3. where there are internal variations both vertically and horizontally within the deposit.

The effects of 1 and 2 are illustrated by the case shown in Figure 1.10. This shows a deposit of fluvioglacial sand, underlain by Jurassic clay, and overlain by boulder clay; the area consisted of a steep-sided hill rising to 103 m in the northern corner of the field being prospected. Thirteen boreholes were made in the field; their records showed a consistent geological pattern, so that this number was deemed adequate to supply the data for assessment.

The first requirement is an adequate topographical survey, from which surface contours at 1-m intervals can be drawn (Figure 1.10(a)). Data from the boreholes are then used to draw contours (related, of course, to the same datum as the surface contours) on the base of the boulder clay. This shows (Figure 1.10(a)) that the base of the boulder clay can be interpreted as a flat surface rising from south to north. The intersection of the surface contours with the basal contours will give the depth* of the boulder clay at that point.

By joining the intersection points (examples of which are ringed in Figure 1.10(a)), Figure 1.10(b), which shows lines of equal depth of boulder clay, can be constructed. It is worth while at this stage also to mark on the map the actual depth of boulder clay recorded in the boreholes, as this provides a useful check and guide in drawing the lines of equal depth. By planimeter measurement, quantities of boulder clay overburden can be calculated. Note that the zero line is the outcrop of the base of the boulder clay, whose position can often also be checked on the ground by surface indications.

The next stage (Figure 1.10(c)) is to draw contours on the base of the sand, which is of course the top of the Lias clay. These show a gently inclined surface falling from 94 m above datum in the south to below 92 m in the north. The depth of sand can be found by using intersection points as before; but it must be remembered that outside the boulder clay outcrop it is the intersection of surface contours and base-sand contours which is used, while within the boulder clay outcrop it is the intersection of base-boulder clay and base-sand contours. Again, ringing of intersection points, and noting actual depths in boreholes are useful techniques.

The result of this stage is Figure 1.10(d), which shows lines of equal depth of sand over the whole field; and the application of planimeter techniques can readily convert these to volumes of sand. At this stage allowances can easily be made for possible limitations on extraction. For instance, if it were decided that boulder clay overburden removal in excess of 2 m was not feasible, the transfer of the 2-m overburden line from Figure 1.10(b) would allow this to be excluded from measurement. If the highway authority were to ask for a 100-m wide safety zone against the road on the side of the property, the effect of this could simply be calculated.

There has never been an objective check of the accuracy of this method, although long experience in the industry suggests that its forecasts have never seriously been faulted. Clearly, the larger the scale used the greater the accuracy; in general the 1:2500 scale, which

* The term thickness is generally used by geologists to describe the total stratigraphical thickness of a deposit, irrespective of any erosion surface which may cut across it, and geological maps which show isopachytes, or thickness contours, usually refer to this parameter. In surface mineral deposits the stratigraphical thickness is usually diminished by erosion, and for this actual, in-the-ground measure it is suggested that 'depth' is the more appropriate. No equivalent term to isopachytes has been used; isobath, though technically correct, has only been used of water.

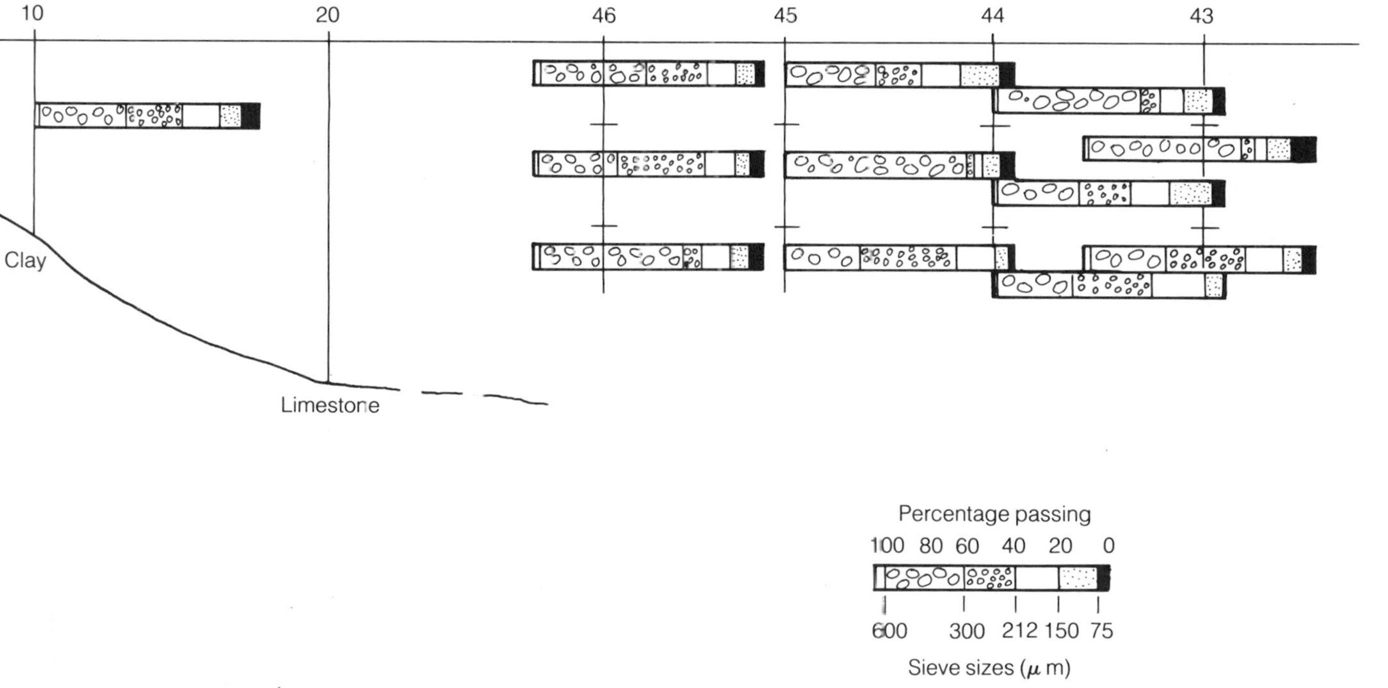

Figure 1.11 Geological section of an alluvial deposit: horizontal scale 1:2500; vertical scale 1:200; 10, 20, 46, etc. indicate boreholes; bars illustrate percentage passing sieve grades.

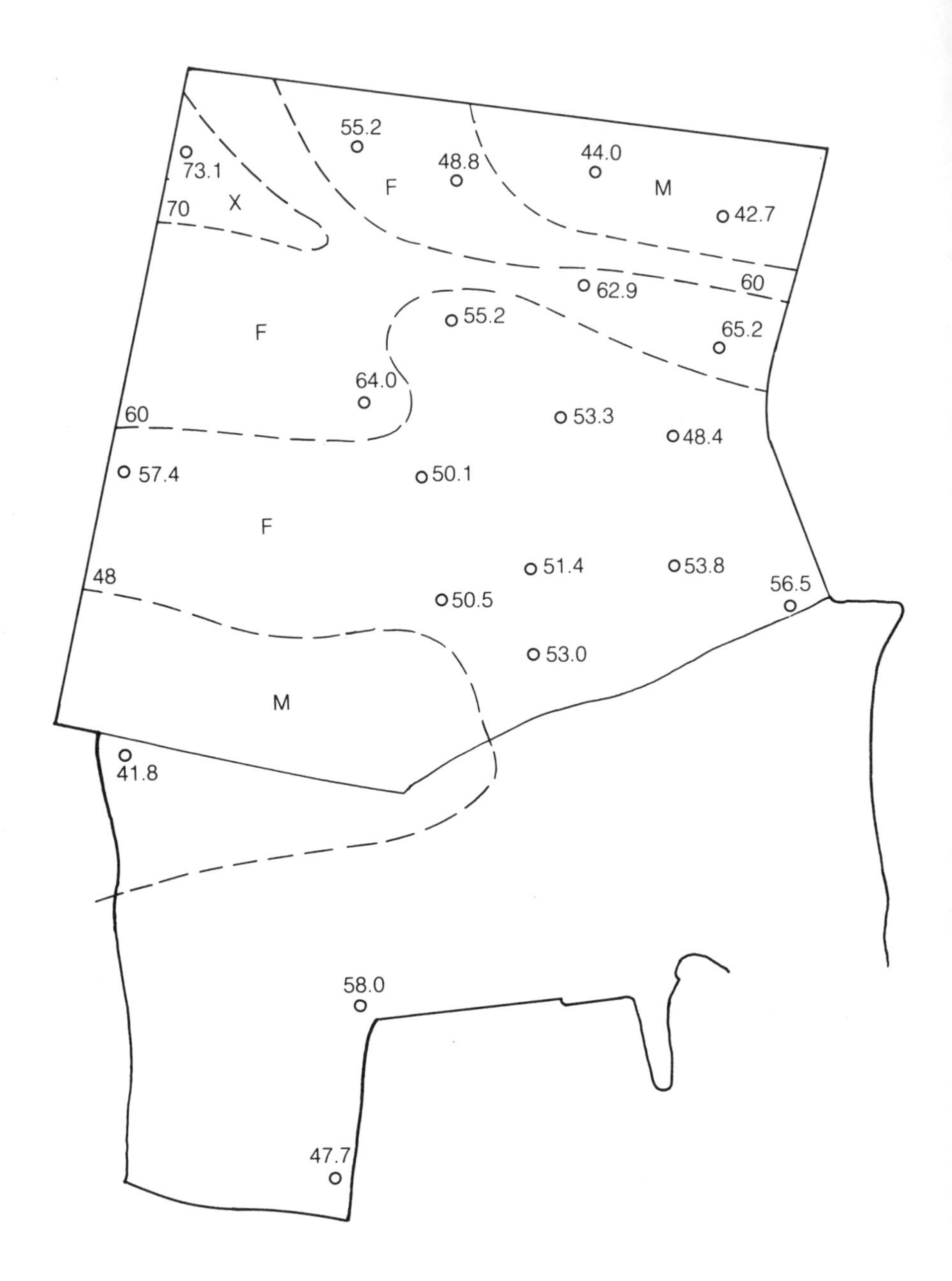

Figure 1.12 Map of an area of alluvial deposit (as Figure 1.10), scale 1:2500, to show percentage passing 300 μm sieve (average for total depth of borehole): ○, borehole position (with percentage passing 300 μm); X, fails BS 882 (>70%); F, satisfies BS 882 for fine concreting sand (<70%:>48%); M, satisfies BS 882 for medium concreting sand (<48%:>40%).

has been used in the examples given here, is adequate for preliminary calculations – i.e. as a basis for decision on whether to proceed or not, or as supporting material for licence or planning applications. If more precision is needed (as it frequently is where it is the basis of financial transactions), scales of 1:1000 or 1:500 are usually the least that is required.

Where there is lateral and vertical variation within a deposit, assessment becomes even more complex, and every type of problem needs its own solution. One example may perhaps suffice (see Figures 1.11 and 1.12). This displays an alluvial sand deposit, which shows very variable sieve analyses in a series of boreholes; the need is to find some kind of pattern within these sieve analyses which will enable the geologist to quantify the different types of sand. The method chosen was to draw up a series of horizontal sections, on the basis of the borehole records in which the sieve analyses are portrayed as bar diagrams. Using this method, any systematic variations – e.g. if there were any consistent changes with depth, or if there were regular variations from one side of the site of the other – would become immediately apparent. In fact, in this instance, no such systematic variations could be identified, the changes appeared to be entirely random. So it was decided to average the results for each borehole, and to map the results (Figure 1.12). A map was drawn for each sieve grade, and Figure 1.12 shows that for 'percentage passing 300 µm' – an important size in the British Standard for sands for concrete (section 4.2), a pattern is at once discerned; moreover it is a pattern which the geologist will recognize as one not unlikely in the deposits of an east–west flowing river, as these are known to be. Moreover, it enables the geologist to identify areas in which the sand lies outside specification (X), and those which fall into categories suitable for fine (F) or medium (M) concrete sand; and, used in conjunction with a map of depth of deposit, these can be readily converted to volumes.

1.3 QUARRY DESIGN AND MANAGEMENT

The exploration stage should have provided the information necessary to build up a three-dimensional picture of the deposit – its lateral extent, its thickness, its geological structure, its internal variation, and the depth and disposition of any overburden. The boreholes should also have been used to provide information on groundwater levels within the site. Using this information, the planning and excavation of the quarry can then proceed on a logical

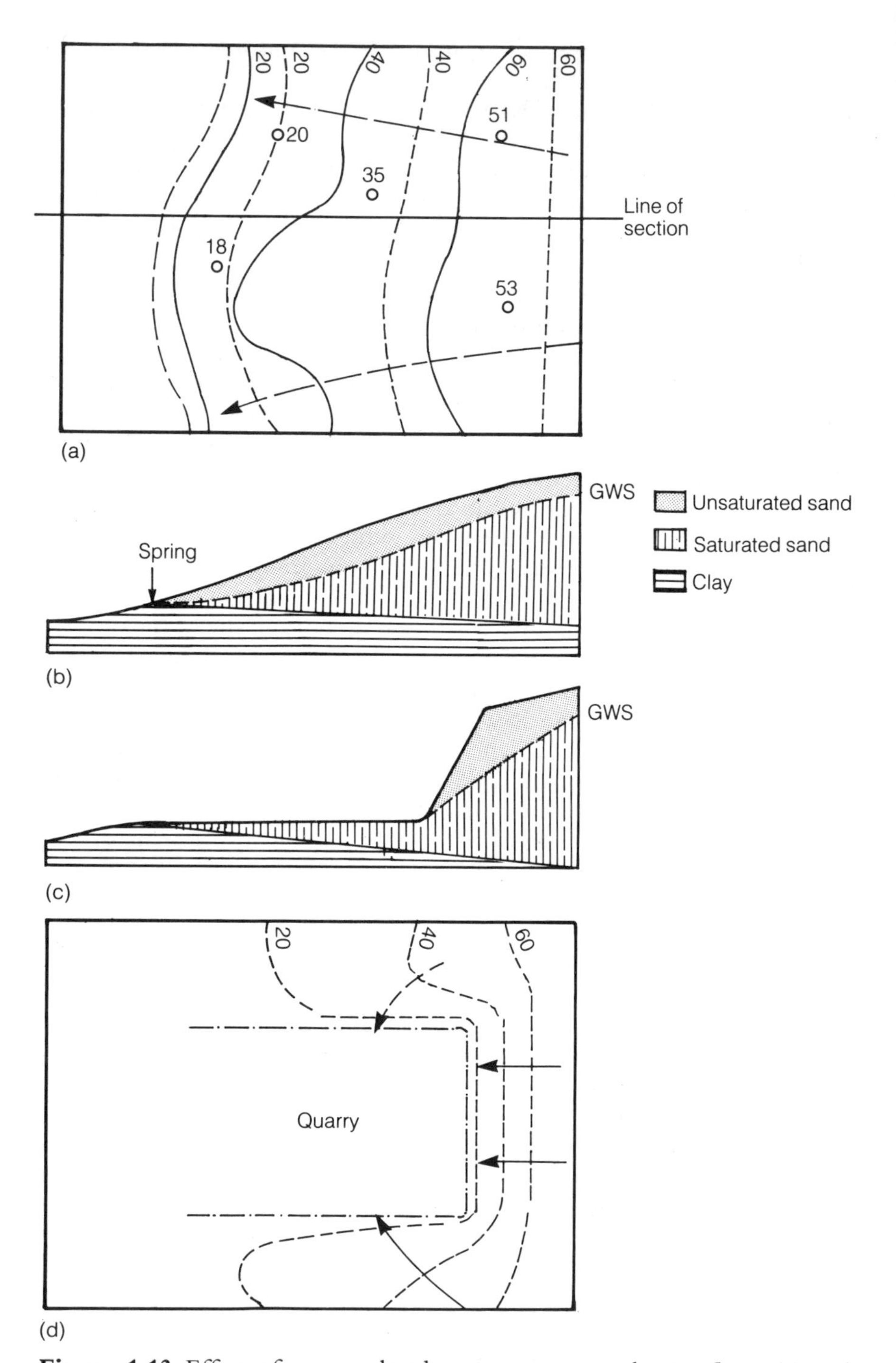

Figure 1.13 Effect of quarry development on groundwater flow: (a) and (b) map and geological section before development; (c) geological section of quarry; (d) groundwater situation after quarry development: ○ piezometer locations, with groundwater surface (GWS) levels: ___20, surface contours; ----, groundwater contours; — — —, sand/clay outcrop; ———→, flow lines of groundwater.

basis. There are five areas in which specific technical knowledge is required; they are

1. the means of excavation, i.e. the machinery, plant and methods needed to excavate the deposit
2. the method of excavation, i.e. the way the deposit has to be worked to ensure a continued supply of the desired product
3. the planned stripping, removal and disposal of topsoil and overburden
4. the methods of dealing with water in the quarry, and
5. ensuring the safety and stability of the quarry faces.

Of the above, 1, 2 and 3 require the preparation of a detailed quarry plan, the basis of which is the three-dimensional interpretation made by the geologist. Items 4 and 5, however, lead into the fields of hydrogeology, soil and rock mechanics. These are specialist fields, and in a large project a geologist might rightly feel the need to seek specialist advice; this is not always available, however, and at least the basic principles should be understood.

Groundwater control and planning is essential in most quarries. If the quarry is in an impermeable clay, or is situated well above the local water-table, groundwater may not be a problem. If the quarry intersects the groundwater table, development of a quarry will change the pattern of water flow. It is necessary to predict the effects of this change for many reasons. The amount of water inflow into the quarry must be known, so that proper provision is made for its escape, or removal by pumping. The effects of the change of flow will undoubtedly extend outside the quarry area, so that there is a need to ensure that any other users of the water are not affected – either by reduction or contamination of their supply.

Figure 1.13 illustrates a simple case; it shows the development of a quarry in a sand horizon overlying an impermeable clay. The clay acts as an aquitard preventing the downward movement of water, so that water can percolate through the aquiperm above, and accumulate as an aquifer (saturated zone) at the base of the sand. The top surface of this saturated zone is known as the water-table or groundwater surface (GWS).

In order to study the groundwater pattern it is necessary to have subsurface data; it is economic if the exploration boreholes can be so used, but if this has not been done, special boreholes will need to be constructed. They are converted to piezometers by inserting a porous base and lining the hole (usually with plastic). Water will rise in these to the groundwater surface, whose level can then be measured. From these levels a contoured map of the groundwater surface can be constructed. This is most informative; the flow direction of the groundwater is at right angles to the contours, and the rate of flow is proportional to the spacing of the contours. Before quarrying (Figure 1.13(a)), the water escapes from springs at the outcrop of the base of the sand; when the quarry is developed (Figure 1.13(d)) the excavation will intersect the groundwater surface, and water will flow into the base of the quarry; at the same time the groundwater surface will be steepened, so the rate of flow will increase.

It is clearly very important to know the volume of water which is likely to enter the pit in this way. A measure of this can be obtained by gauging the flow from the springs; if this is large, it will be necessary to obtain an accurate estimate. The rate of flow through permeable strata is determined by the inclination of the groundwater surface, and by the ability of the rock of transmit water – its transmissivity.

The inclination can be measured from the predicted groundwater surface contours; the transmissivity can be derived from laboratory measurements on the permeability of samples, or, more satisfactorily, from pumping tests in the field. This latter method involves the pumping out of a measured flow of water from a central borehole. This creates a cone of depression in the groundwater surface around the pumped borehole, which extends outwards until it intersects other boreholes, where it can be measured as the drawdown. From the extent of the drawdown the transmissivity of the strata can be estimated, which will then permit the calculation of the water flow. While this appears a complex procedure, it can be carried out quite simply in the field; and there are also somewhat simpler pumping-in tests which will give satisfactory results.

The exact position of the groundwater surface will vary with seasonal conditions; at times of high rainfall, the recharge of the aquifer will exceed the discharge from the springs, and the groundwater surface will rise; in times of drought it will fall. So it is important that observations of the groundwater levels in the piezometers should be continued over as long a period as possible, so that the best and worst conditions can be ascertained.

The possible effects of the quarry on neighbouring land, and particularly of nearby water usage, can be assessed by extending the observation network so that the flow pattern can be observed. The amount of drawdown recorded during the pumping experiments can, by extrapolation, be used to determine how much any neighbouring source is likely to be derogated by the proposed development.

The design of a quarry, and establishment of a pattern of working which assures that the quarry face shall be stable at all times, must also depend upon an understanding of the geology. The study of the stability of natural and artificial slopes is the remit of the twin specialisms of soil mechanics and rock mechanics, which form a major part of the science of engineering geology. Engineers make no distinction between true soil (in the agricultural sense) and unconsolidated rock; so that soil mechanics deals equally with true soil, the weathered zone of consolidated rocks, and unconsolidated rock; while rock mechanics principles apply only to consolidated rock masses.

Soil mechanics principles are therefore applicable to quarries developed for clay, sand and gravel, while only quarries which are a source of rock for construction stone or crushed rock aggregate can be studied by rock mechanics methods.

Soil mechanics treats a rock mass as a uniform entity, and defines parameters which determine the behaviour of that entity under different conditions. A basic parameter is that of **shear strength** – simply defined as the ability of a rock mass to resist shearing forces applied to it, either by loading (as in foundations for buildings etc.) or by gravity (as in a quarry face). Shear strength is regarded as the product of two variables, both determined by the characters of the rock itself – these are the **angle of internal friction** (φ) and **cohesion** (c). Cohesion is mainly dependent upon the presence of clay minerals, and so is at a maximum in argillaceous sediments, while virtually absent in sands and gravels. Some degree of cementation may have the same effect as cohesion, while cohesion may be much reduced by the presence of small-scale discontinuities. The angle of internal friction is a parameter which can be measured in the laboratory on specially collected samples.

Using these two measures (c and φ) it is possible to calculate the angle at which a slope in the particular rock mass (soil) will be unstable. If cohesion is zero, the slope cannot stand at an angle greater than φ; higher values of c allow the slope to stand at higher angles. Stability is decreased by the presence of water, measures of which are introduced into the equations. These equations relate stability to a factor of safety; while it might be generally agreed that a factor of safety of 1 is highly dangerous, while a value of 10 might be very safe, there is no wholly objective means of deciding the appropriate factor of safety to adopt; what might not be allowable for a major motorway cutting, for example, might be feasible for a small, temporary quarry.

The basis of rock mechanics is entirely different. While the strength of the rock is one factor, and a value for φ can be calculated which is related to the unconfined compressive strength (section

2.1) of the rock (Figure 1.14), it is recognized that the major cause of instability lies in the discontinuities (bedding, jointing, etc.).

There are a number of different ways in which a rock slope may fail, and these are related to the disposition of the discontinuities (Figure 1.15). Plane failure (a) occurs when the slope parallels a major discontinuity – a major joint or bedding plane – and blocks slide

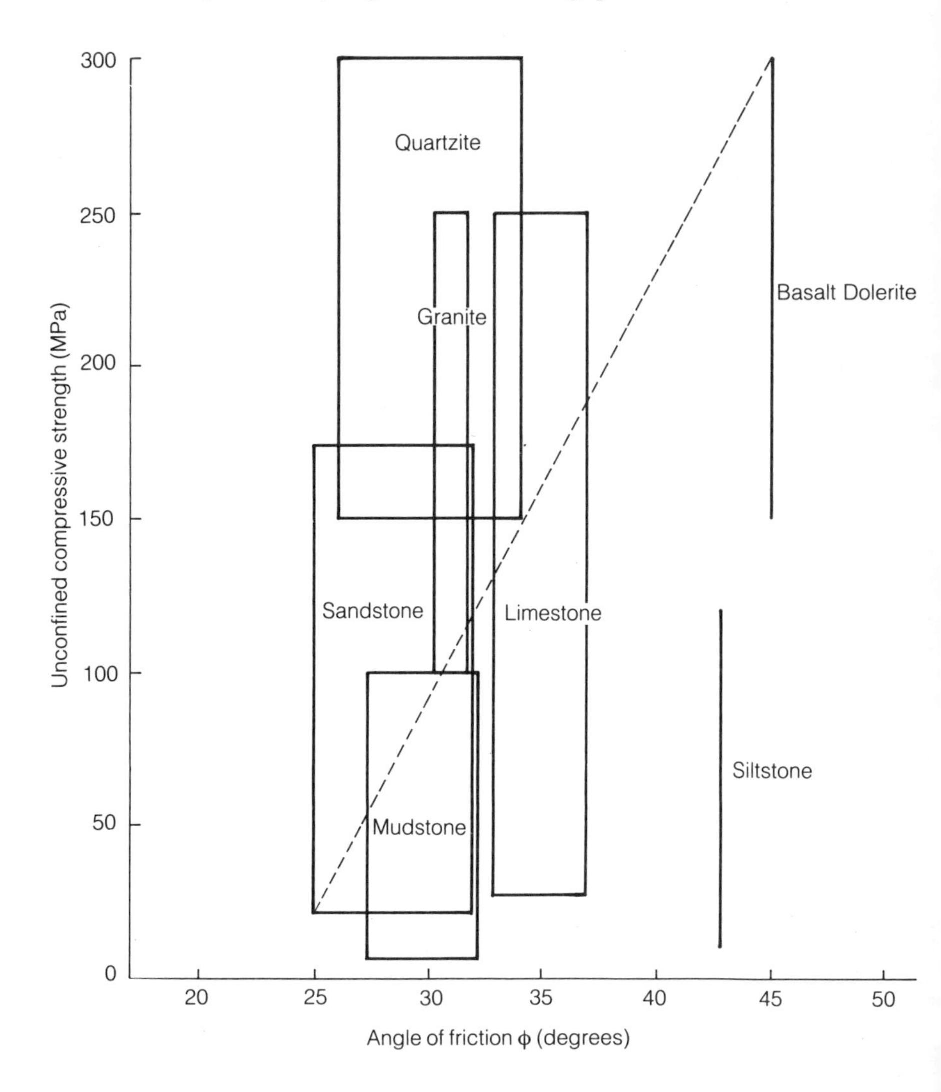

Figure 1.14 Relation between angle of frictional resistance (φ) and unconfined compressive strength in various rocks. Data from McLean and Gribble (1979).

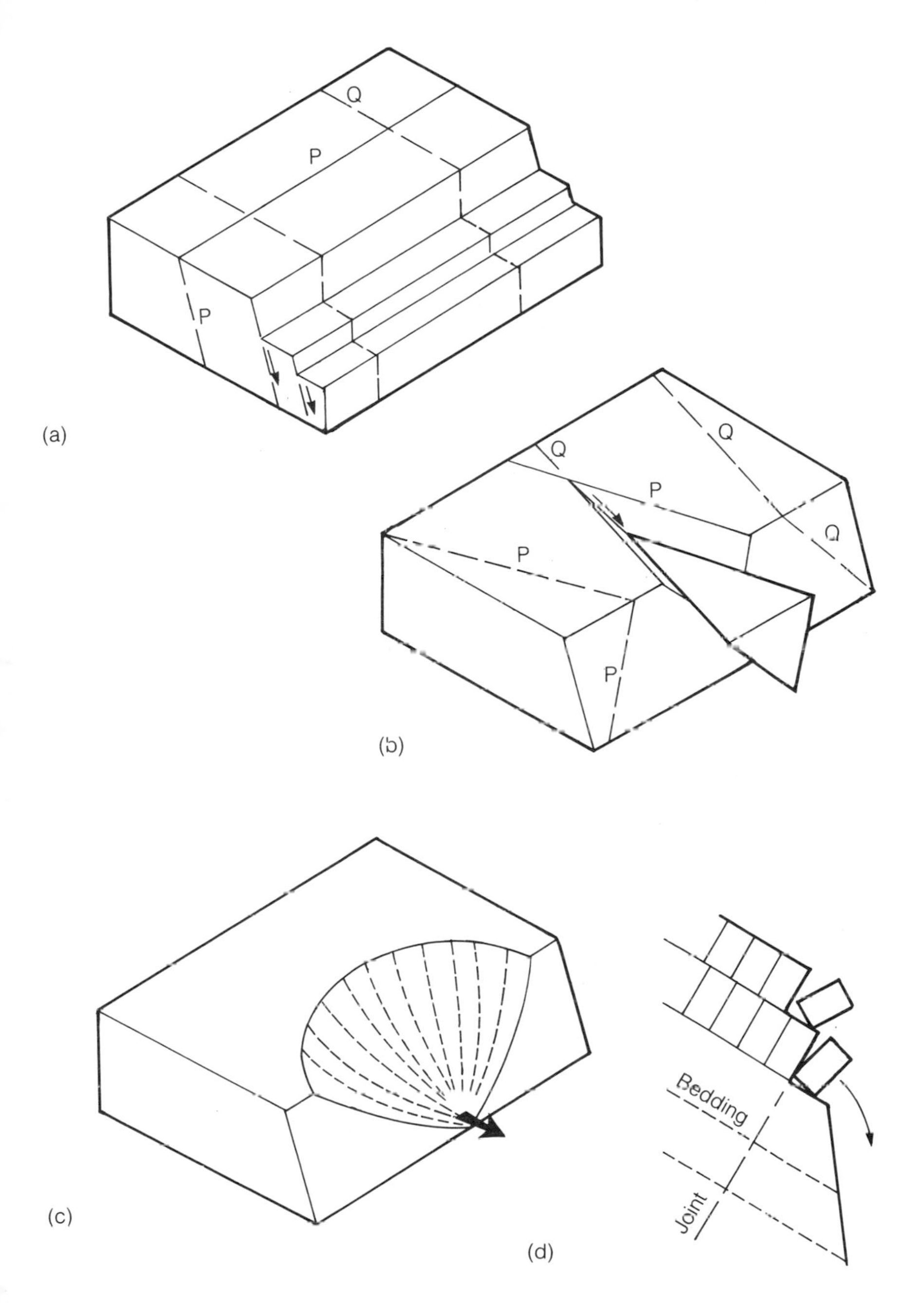

Figure 1.15 Slope failure in consolidated rocks: (a) plane failure; (b) wedge failure; (d) block toppling; (c) circular failure. P and Q, discontinuities; ———→ direction of movement.

down along this plane. Wedge failure (b) occurs when two joints intersect, and wedge-shaped blocks slide forward out of the face. In these two cases failure occurs when the angle of the plane of movement with the horizontal – i.e. the plane itself in (a), or the intersection line in (b) – exceeds the angle of frictional resistance. Failure occurs at lower angles if the surface is lubricated by water, or by softer infills, and at higher angles when the surface is rough. Failure by block toppling (c) is a different mechanism, and the angle at which it occurs is related to the size and shape of the blocks involved. Finally, circular failure (d) occurs when fracture spacing is so frequent that the rock behaves like an engineering soil, and a rotational slip occurs throughout the rock mass.

The effect of weathering on both soil and rock face stability is extremely important. It must be recognized that the physical properties of the weathered zone near the surface are probably quite different from those of the intact rock at depth. Nor must the longer-term effects of exposure of a quarry face be overlooked; a pit dug into a cohesive soil may be stable at quite high angles, but if this is left to weather, the development of multiple discontinuities will reduce this angle dramatically. While there are ways of palliating this – by reducing water flow, and encouraging the growth of vegetation – substantial slope-failures are difficult to stop once they have begun. The design and planning of a quarry should therefore look not only at its working life, but at its long-term future and restoration.

The above is given as a general explanation of principles – but hydrogeology, soil and rock mechanics are complex specializations whose full understanding does not lie within the training of most 'general practitioner' geologists. If water conditions or face stability are likely to present problems, and particularly if their solution may have large financial implications, the geologist should not hesitate to take expert advice in these fields.

2

Construction stone

Quarried stone is used in a variety of ways in the building and construction industry; shaped and sized as dimension stone, or roughly broken and used for rip-rap or rock-fill. When broken further into gravel or sand-sized pieces, and used as a primary ingredient in the production of concrete or tarmacadam, it becomes known as aggregate.

2.1 DIMENSION STONE

Stone which has been cut and dressed into regularly shaped pieces is known as dimension stone (**ashlar** to the architect); the dimensions can range from several metres in three directions as in the large blocks used for public buildings, to thin slabs used for ornamental cladding or for floor tiles.

The desirable features of a good dimension stone are

1. structural strength
2. durability
3. attractive appearance, and
4. ease of quarrying and dressing

while other factors which determine the choice of stone to be used are

5. availability (i.e. ease of transport from quarry to site), and
6. environmental consequences of quarrying.

The ability to carry a required load without failure is determined by the structural strength of the rock. In the case of large structural blocks, which are needed to support a building, compressive strength is the most important, while in the case of floor tiles a high tensile strength is required. Strength in a block of rock is determined by two factors. First by its mineralogy; some minerals (quartz for example) are inherently strong; others (for instance clays and micas) are weak. Secondly by the presence or absence of discontinuities; in sedimentary rocks these will be bedding and jointing; in metamorphic and igneous rocks cleavage, foliation, etc. If these discontinuities are closely spaced, as they might be, for instance, in a thinly bedded limestone, such a rock is unlikely to be useful as a dimension stone. At the same time, the presence of discontinuities is extremely important in facilitating the extraction of the blocks from a quarry. The ideal situation is thus one in which the spacing of these discontinuities coincides with the size of the blocks which are to be extracted – a fortunate coincidence which is rarely realized.

Although compressive strength, as measured in the laboratory, is an acceptable measure of the ability of a rock to support a load, it does not take account of the slow changes which can occur within a rock structure if placed under heavy loads for a long time. These changes are well known to structural petrologists, who recognize the effectiveness of pressure solution along points of contact of individual grains of quartz or calcite, thus allowing the rock to be deformed without obvious fracture. Such a mechanism probably accounts for the bowed form of slender pillars which is not infrequently seen in our major cathedrals as, for example, in the Purbeck marble pillars in the nave of Salisbury cathedral.

Durability is the ability of a rock to resist exposure to its environment – which may be the atmosphere, where the rock is used as a facing for a building, or groundwater where used for foundations, or the sea in marine situations. Geologists have long studied the weathering processes on natural rock outcrops, and from this we can recognize the major role of interstitial water in this process. The crystallization of such water, either by freezing, or by the formation of crystals of salts dissolved in it, exerts a pressure within the rock which is readily able to destroy the whole structure. These dissolved salts can originate from chemical weathering of the rock itself – as, for instance when pyrite decays to sulphates; from materials used in the mortar; or from saline groundwaters rising from below. The process is accelerated in the polluted atmospheres of cities. While clearly of the greatest significance in humid climates, and where freeze-and-thaw is a common phenomenon, there are few climates so arid that this process is not effective. Thus, for example, the cathedral of Cadiz, in southern Spain, enjoying an exceptionally dry climate, has suffered almost total disintegration of its stone-carvings by the activity of rising saline groundwater.

It follows that porosity is a major consideration, and much effort has been devoted to devising means of measuring this, as a predictor of stone durability. It has been noted that the behaviour of pores is related to their size, those below 5 μm retaining their water even under applied suction. There is a widely held belief that this microporosity is of particular significance in rock durability.

There are various methods of assessing the porosity of a rock sample in the laboratory, set out in Figure 2.1. They are based on the comparison of the dry weight of the specimen, and its weight when soaked in water under various conditions. Porosity percentage is a measure of the fully saturated rock, while the saturation coefficient measures the water content after total immersion for 24 hours. Microporosity is a measure of the water content retained after a fully soaked sample has been subjected to a negative pressure – it is believed to measure the volume of pores which are less than 5 μm in diameter.

Capillarity is measured by placing the specimen with one end in water, and weighing it after a period of time has elapsed. It has been observed that the process of water uptake in many rocks is a two-stage process (Figure 2.2); a period of rapid uptake is followed by a stage of much slower rate. The Belgian standard for building stones uses these uptake rates (designated as α_1 and α_2 respectively) to define

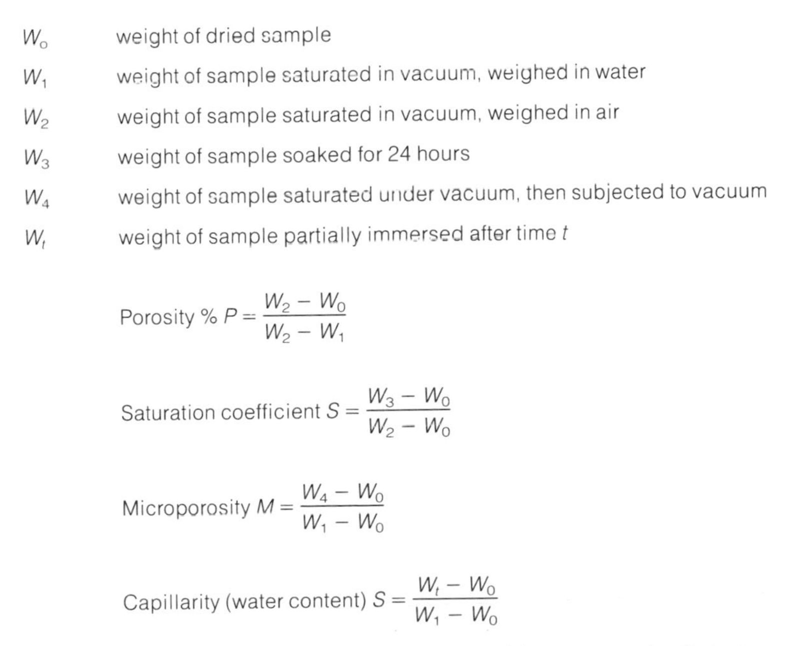

Figure 2.1 Measures of porosity used for building stones in Britain.

Belgium: Standard NBN B 05–201

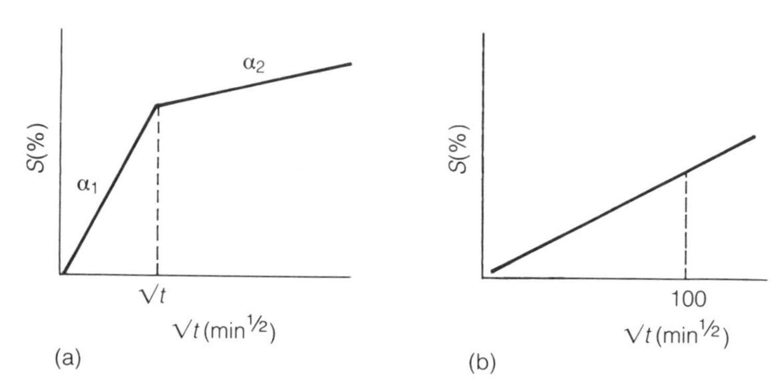

Figure 2.2 Calculation of capillarity for Belgian and French building stone standards.

two coefficients – G, which is based on α_2, and GC, based on α_1. In some rocks the two-stage uptake does not occur, and here the uptake after 100 minutes is used in these calculations. The French standard uses a similar measure based on the total water absorbed per unit area.

In Britain, a further test has been devised – the crystallization test. This attempts to simulate the effect of ice in the pores of a rock by repeated soaking in a concentrated solution of sodium sulphate, followed by drying. An actual freeze–thaw technique is used in France to define this characteristic.

In France, Belgium and Britain, the results of the various tests, in which those for porosity predominate, are used to place each building stone in a small number of classes (four in France, five in Belgium and six in Britain), which are then used to define the types of climate, and the position in a building, in which the stone may confidently be used. Thus in the British system, a limestone of Class C (which would include many Jurassic limestones) would be recommended for use for exposed copings and parapets at an inland site with no pollution, but only for more protected positions in areas of high pollution, or at coastal sites; for plain vertical walls only at coastal sites subject to frost; and not at all where a coastal site is subject to high pollution.

Chemical attack of building stone by atmospheric waters is a phenomenon much intensified in the polluted air of modern industrial cities. Limestones are the most affected, since calcite is readily soluble in dilute acids, and most limestones which have been exposed to such atmospheres for only a few decades display a deeply etched surface. In London, the stone chosen for the construction of the new Houses of Parliament in 1840 was Permian magnesian limestone – a wise choice, since dolomite is much less readily soluble than calcite. Many sandstones have a calcareous cement binding the quartz grains together, and so are equally susceptible to attack. Testing of samples for chemical resistance is carried out by a simple acid-immersion technique.

A further hazard, particularly in hot climates, is the change of some silicates found particularly in basic igneous rocks, into expanding clay minerals – examples of rock decay from this cause have been reported from Spain and Italy. This process has also seriously affected the basalt used extensively for dyke protection in the Netherlands; in order to measure the susceptibility of such rocks in this respect, a test has been devised in which the rock is milled for a standard period of time, and the amount of smectite developed measured by X-ray diffraction (Figure 2.3).

The compressive strength of rocks used for construction varies enormously (Chapter 3, pp. 69–72 and Figure 3.1), from <10 MPa for some Mesozoic limestones, to >260 MPa for some granites. Experience suggests that compressive strength is not normally, of itself, of great significance in the assessment of building-stone quality. Compressive strength ought, however, to have a relationship to pore volume, since the presence of pores ought to weaken the structure. Some such relationship (Figure 2.4) can be seen in sandstones, although the scatter of points on the graph is wide enough to show that other factors must be involved. The scatter is even wider on the graph of limestones (Figure 2.5); since these must be essentially monomineralic this is somewhat surprising, but is probably related to grain size and differing methods of cementation. Similar graphs for microporosity and saturation coefficient against compressive strength also show a lack of correlation (Figure 2.6), showing that neither the smaller (<5 μm) nor the larger pores have any close relationship to rock strength. The exception is in microporosity, where the concentration of points in one-half of the graph suggests that this factor sets a limit to the strength, e.g. no rock with >60% microporosity has a greater strength than 140 MPa.

As might be expected, a degree of correlation between the various measures of porosity can be established. The crystallization test, which might be thought to represent the nearest approximation to the real weathering process, shows a moderately good correlation with the saturation coefficients, but none whatever to porosity

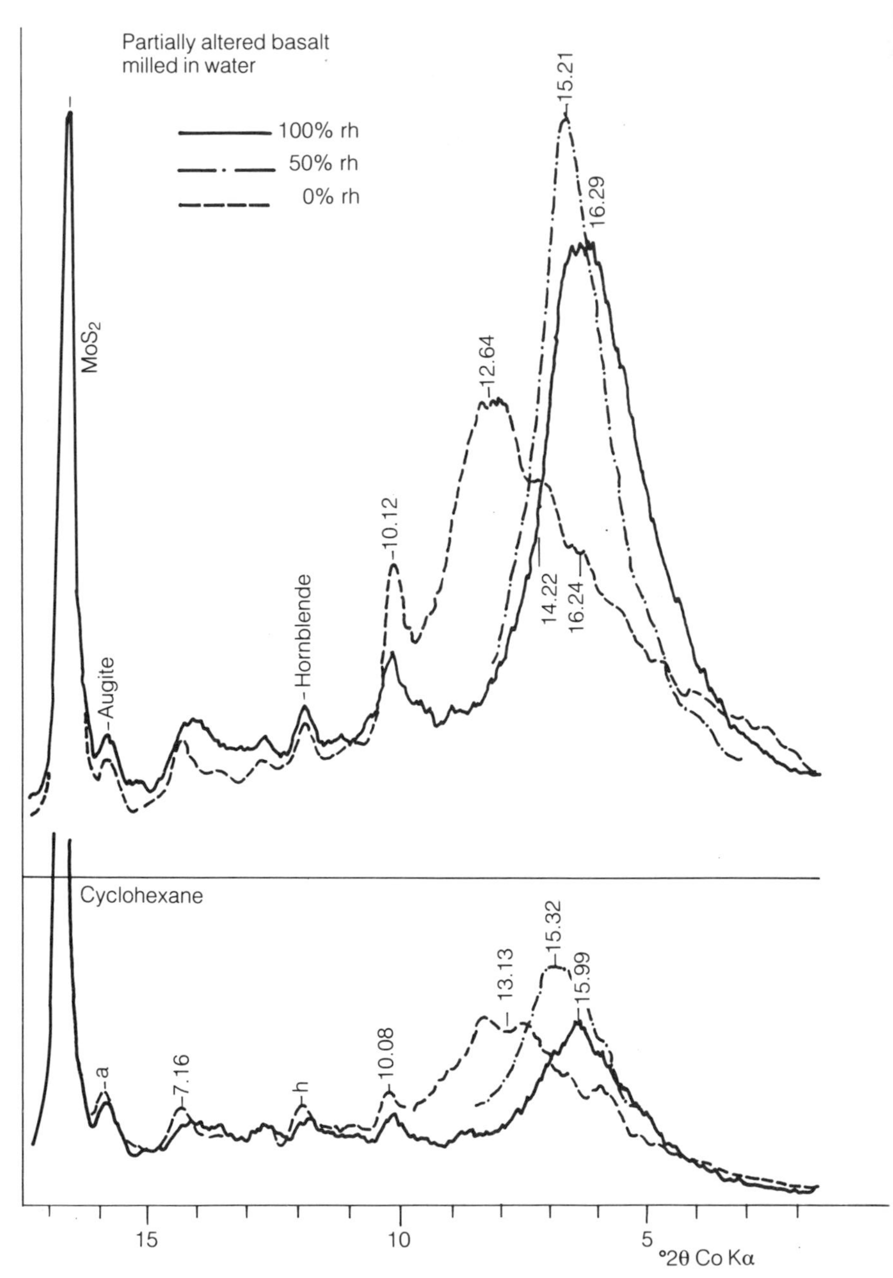

Figure 2.3 X-ray diffractograms of basalt ground in water, and in cyclohexane. Note that the height of the 14 to 17 Å peak (which indicates the amount of smectite present) increases when milled in water, but not when milled in the anhydrous cyclohexane. rh, relative humidity. After Kühnel and van der Gaast (1987).

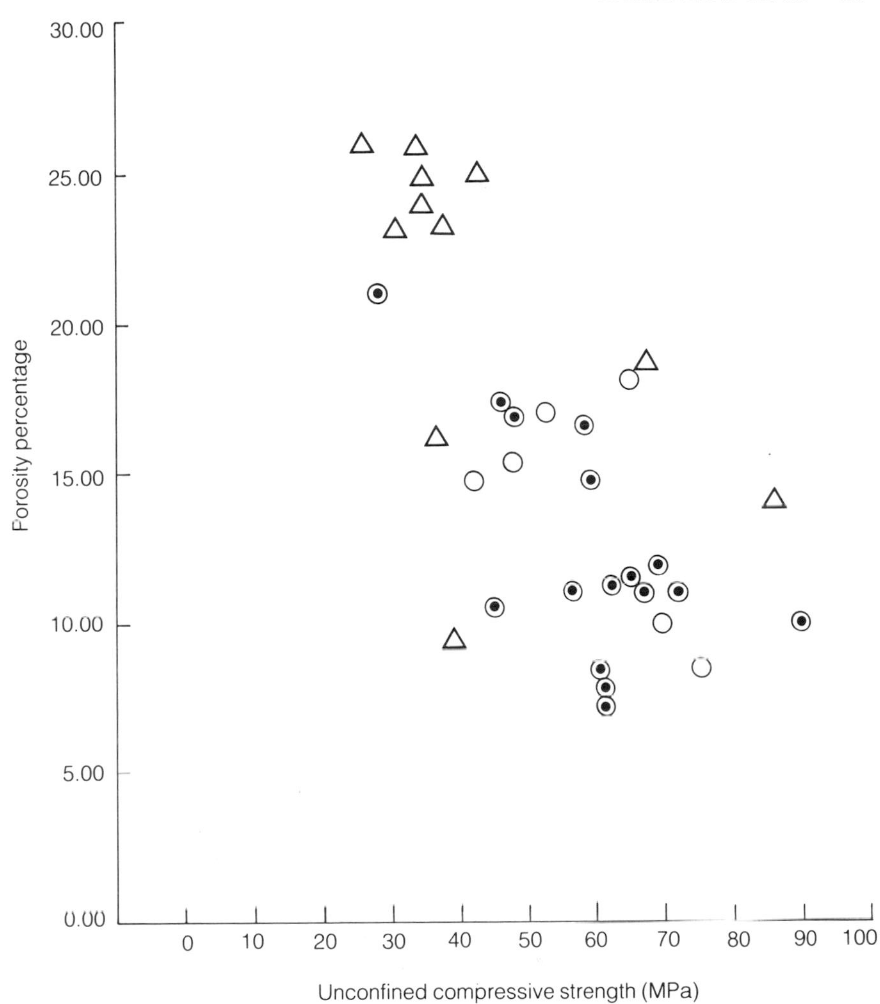

Figure 2.4 Relationship between unconfined compressive strength and porosity percentage for British sandstones used in buildings. ○, Carboniferous (Millstone Grit); ⊙, Carboniferous (coal measures); △, Permo-Triassic.

percentage and microporosity percentage (Figure 2.7). Nor would the crystallization test, as used in Britain, be adequate to place building stones in the French and Belgian classificatory zones.

The conclusion must be that none of the existing tests adequately measures the properties of a building stone which determine its durability. In addition to mineralogy and pore structure, these

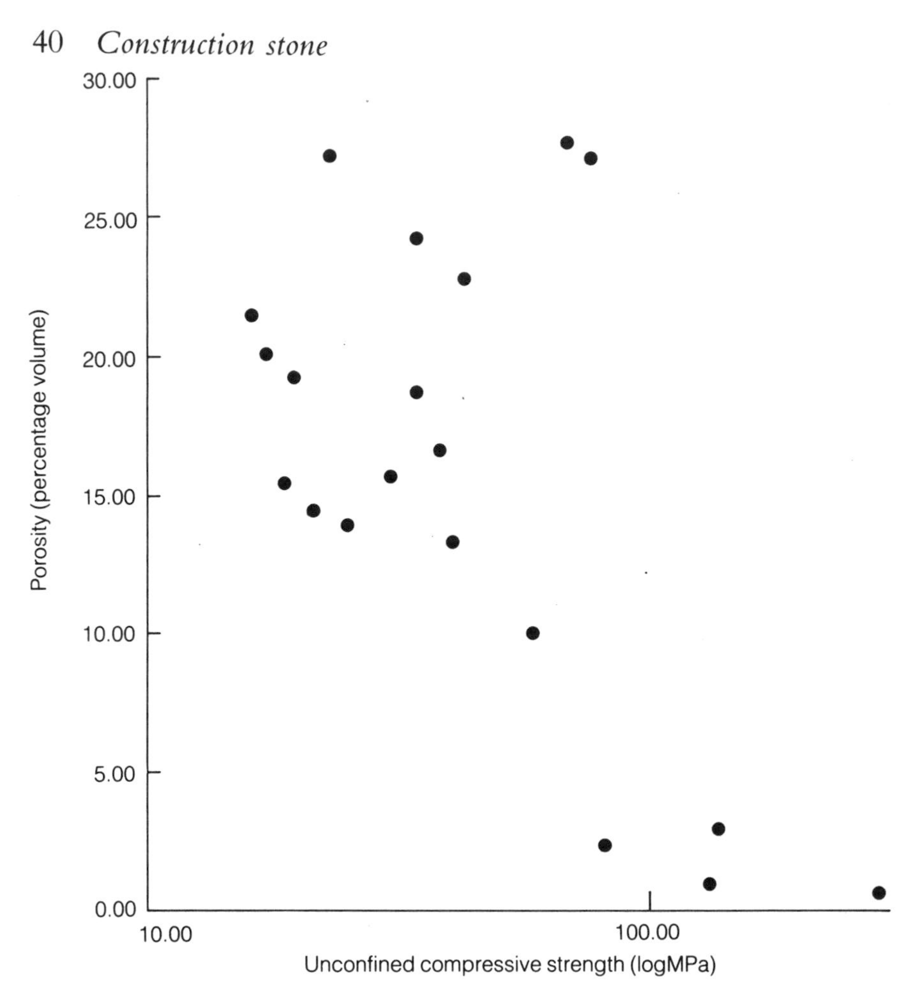

Figure 2.5 Relationship between unconfined compressive strength and porosity percentage for British limestones used for building.

factors are probably particle size, texture and cementation, none of which have been studied in this connection. The present system of tests, although adequate to reject wholly unsuitable rocks, and able to divide others into broad classes, is as yet inadequate to predict with any accuracy the durability of a particular building stone.

2.2 HISTORY OF DIMENSION STONE

Natural stone must surely be one of the oldest building materials used by man. The people who erected the rows of megaliths at

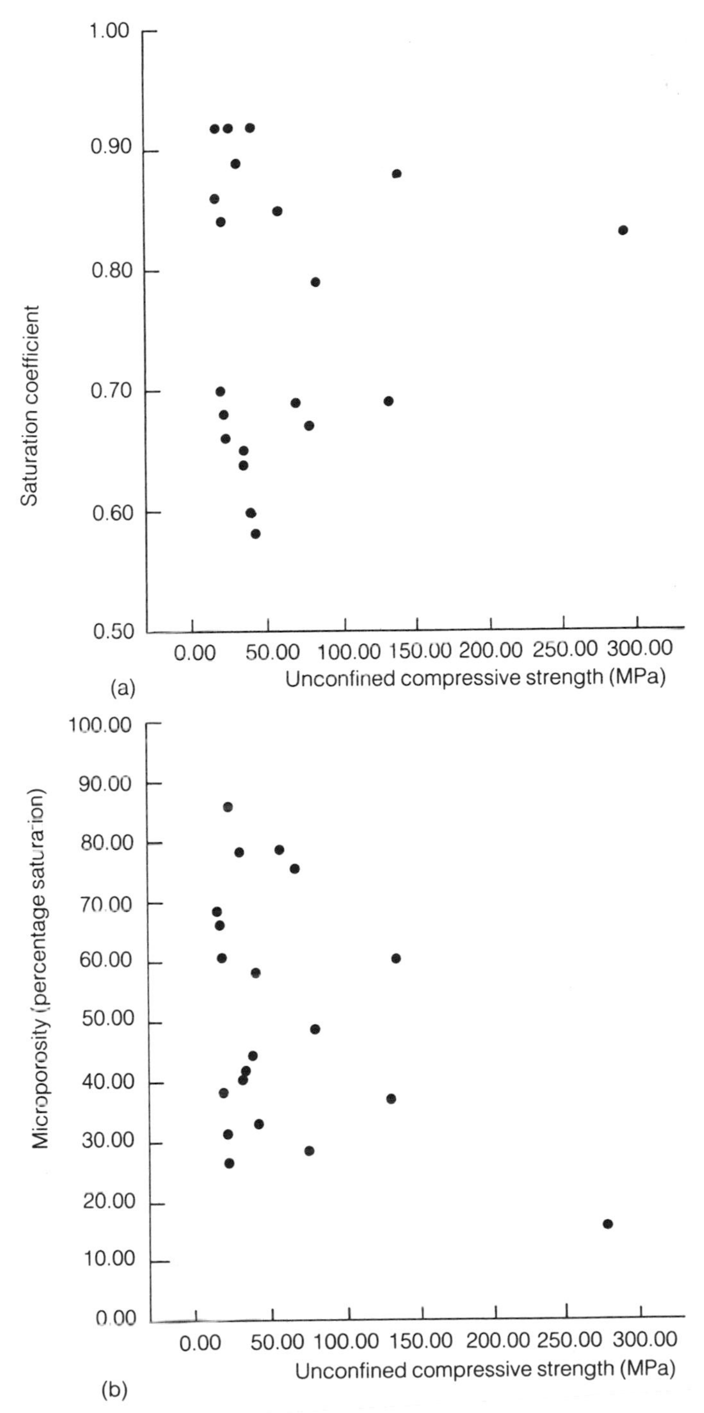

Figure 2.6 Relationship between unconfined compressive strength and (a) saturation coefficient and (b) microporosity percentage, for British limestones used for building.

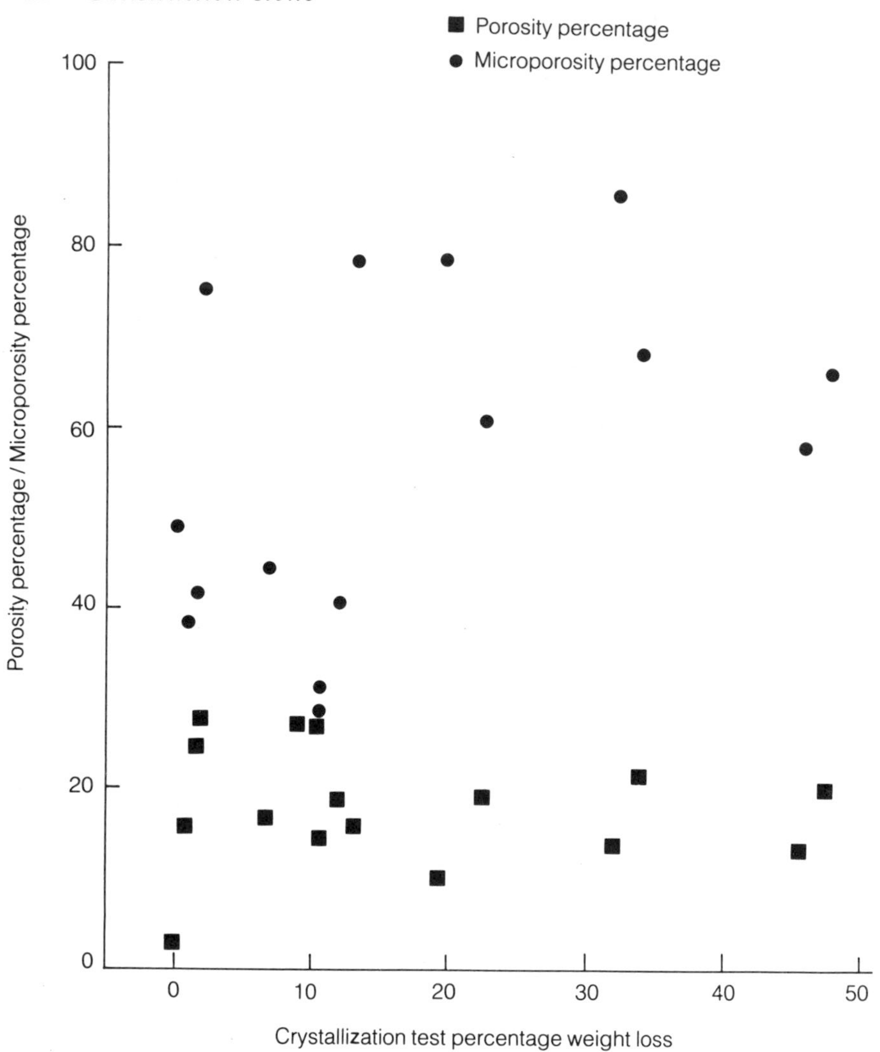

Figure 2.7 Relationship of porosity percentage and microporosity percentage to results of crystallization test, for British limestones used for building.

Carnac in Brittany, around 2000 BC, used mainly relict blocks weathered out of the surface of the local granite; some of these blocks show evidence of having been prised away along the conspicuous joint-planes, but little attempt was made to shape them. A little later (from *c.*1800 BC) the builders of Stonehenge used sarsens – massive blocks of a Tertiary silcrete found abundantly in the weathered surface of the Wiltshire downs – with little or no shaping. The Mycenaean civilization of Greece (*c.*1500 BC) perfected the Cyclopean style of construction, in which massive, but untrimmed, blocks were fitted neatly together, as in a giant dry-stone wall. The ancient Egyptians were more advanced; working with the much softer Tertiary limestones, they produced shaped blocks as early as the First Dynasty (*c.*3000 BC), although effective shaping of granite in Egypt was not achieved for another thousand years. Later Greek and Roman architecture shows, of course, the high level of stone-dressing that had been achieved by those civilizations; and since that time stone buildings have figured prominently in both the 'polite' and the 'vernacular' architecture of every country in the world.

Perhaps the most extensive use of stone in buildings was in the late nineteenth century, when the wealth of Europe and North America expressed itself in grandiose public buildings, and private opulence demanded great houses and mansions of stone. Today the picture is different. Stone quarrying is both labour and energy intensive, so that the relative cost of building in stone is much enhanced, and the dimension stone industry is now much reduced. Inevitably, it has become concentrated in those areas where geological conditions operate favourably to minimize the cost of extraction. Increasingly, stone has become the symbol of prestige, and it is not surprising that it finds favour with the banks and financial houses of today, as it did with the church builders of medieval times, in conveying in a building a sense of permanence and solidity.

2.3 EXTRACTION OF DIMENSION STONE

In the quarrying process the aim is to produce large rectangular blocks as free of imperfections as possible. Sometimes natural discontinuities can be used. For instance, some of the Jurassic limestones of Britain and northern France have a horizontal bedding combined with two sets of vertical joint-planes, thus dividing the rock into natural rectangular units; and driving wedges into these discontinuities is sufficient to dislodge the block. More often, however, it is necessary to drill lines of closely spaced holes; these are drilled vertically and horizontally on all sides of the block using a percussion drill, jack-hammer or diamond rotary drill; the block is then split away by hammered-in wedges. A mild explosive charge ('black powder') may also be used at this stage. In certain rocks a kerosene-fuelled flame-cutter is used to create the primary slit; this depends for its cutting action on the stress created by the differential expansion of the different minerals in the rocks – thus it is highly successful in granites, but less successful in monomineralic rocks such as limestones (Figure 2.8).

Figure 2.8 Jet-cutting of granite, Elburton Quarry, Georgia, USA.

Another extraction method, much used in the past, for example in the 'marble' quarries of southern Belgium, has used a continuous wire process to cut large blocks from the rock face. In this method a borehole is made vertically from the surface, and another drilled horizontally from the rock face to intersect it (Figure 2.9). A continuous wire is fed through these holes; a sand and water mixture is applied to the wire, which then gradually saws through the rock. Then another pair of boreholes is drilled further along the face, and wire-cut in the same fashion. When the two sides are cut, the block is cut away from base and back in the same way.

A large block, thus freed from the quarry face, is then transported to the finishing plant for slabbing and surface polishing. Slabbing is usually performed by gang saws – for soft rocks sand-armed steel, but for harder rocks diamond saws are necessary – and a variety of finishes, from simple trimming to full polishing, is applied. From the above it will be clear that the production of a finished piece of dimension stone is a costly business, in terms of labour, energy and skill. It is not therefore surprising that the industry has seen a contraction in recent years, and that many quarries formerly producing cut stone have turned to the more readily mechanized, and therefore more profitable, production of aggregates.

Figure 2.9 Continuous wire-cutting of marble; Hellshire Marble Quarry, Jamaica (photograph: H. P. Prentice).

2.4 COMMERCIAL PRODUCTION OF DIMENSION STONE

Small quarries, from which roughly trimmed stone for walls and houses is extracted for local use, are found wherever there is any moderately hard stone. But large-scale commercial quarries, supplying cut stone to a wide and extensive market, are few and far between. Capital expenditure and running costs of such a plant are clearly high, and because of the high cost of the finished product, the market is undoubtedly limited. The producer thus needs an assured supply of consistent material; and large deposits of such suitable material are geologically rare.

The geological circumstances in which large masses of unfractured rock are found are somewhat limited. In sedimentary rocks they are realized when there is continuous and uninterrupted deposition, so that bedding planes are widely separated, and where there has been little subsequent tectonism. In igneous rocks such circumstances are only found where they have been intruded subsequent to a major tectonic phase – as is the case with large granite batholiths. Substantial masses of metamorphic rocks only occur when the tectonic effects have either been very slight, or so intense that almost complete remelting has taken place. Because of the comparative ease with which calcareous rocks can be recrystallized, many useful building stones are marbles, which can be found in areas of comparatively low-grade metamorphism; by contrast completely welded rocks of granitic composition are only found where very high-grade metamorphism has converted them to gneisses.

Thus, in general, most large-scale commercial operations in this field lie in

1. limestones and marble
2. granitic or similar intrusions, or
3. metamorphic basement complexes.

Calcareous rocks – limestone and marble – are particularly useful as dimension stone. Calcite is a relatively soft mineral, so that the cost of cutting and polishing is relatively low; at the same time it is able to take a high polish. Its rhombohedral cleavage assures that the rock may be cut equally easily in all directions, which makes it a great favourite with sculptors; while the lack of elongation in the crystal habit means that, even when strongly tectonized, there is no assumption of linear structure within the rock. Depositional conditions are also helpful; immediately after deposition diagenetic processes take place which involve solution and redeposition, so that even before burial the rock may well contain a substantial amount of wholly crystalline calcite. This process also tends to eliminate bedding discontinuities, which thus only occur when there are major changes of depositional conditions – themselves unusual in the normally quiet waters in which limestone has accumulated.

The ready solubility of calcite in dilute acids makes limestone a less-than-perfect material for use in the polluted atmosphere of modern cities; in these circumstances the surface is often found to be deeply etched after a few years. A percentage of magnesium in the mineral renders the rock less soluble – the choice, therefore, of Permian dolomitic limestone for the Houses of Parliament in London, for instance, can now be seen to have been wise. However, exposure to less aggressive atmosphere can be beneficial, because the slow solution and redeposition of the calcite in the surface layers creates a hard patina which resists further attack by weather. As a flooring material calcitic rocks will not long resist the abrasion of heavily trafficked areas because of the inherent softness of the mineral; despite this they are nevertheless much used in houses and public buildings.

Limestones also tend to be porous (p. 37 and Figure 2.10); geologically old limestones, such as the Carboniferous limestone (Figure 2.10 (a)) may be very impervious, because the extensive diagenesis and lithification which they have undergone has filled the pores with redeposited calcite; but this also makes them hard and difficult to quarry. Younger limestones (Figure 2.10 (b) and (c)) have higher porosities, and thus lower durability.

The industry frequently fails to distinguish the ornamental limestones (which are wholly sedimentary in origin) from their metamorphic equivalents, marketing them both as marbles. Perhaps (to modify Dr Johnson) lapidarists, like their inscriptions, are not necessarily on oath!

In Europe, the Mediterranean area has the most extensive calcareous rocks (Figure 2.11). Italy ranks first amongst the marble producers of the world, with a tradition going back to classical times. Geologically, in early Mesozoic times thick deposits of limestone accumulated in the warm waters of the precursor of the Mediterranean Sea, the Tethyan Ocean. As Africa moved towards Europe, these deposits became involved in the fringes of the Alpine orogeny – but the Appenine folding was much less intense, and involved much shallower burial, than that of the Alps themselves. Thus these limestones were recrystallized to marble, but not strongly fractured. The original limestones were very pure, so that the much-prized pure white marble is found; but traces of iron, chrome and manganese also generate a variety of red, green and brown colorations. Production is centred on the Apuan Alps and in Puglia, both areas where major anticlinal structures have brought the Mesozoic rocks up within the otherwise ubiquitous Tertiary cover. Smaller operations are found in Venetia and Sicily – both producing a wide variety of

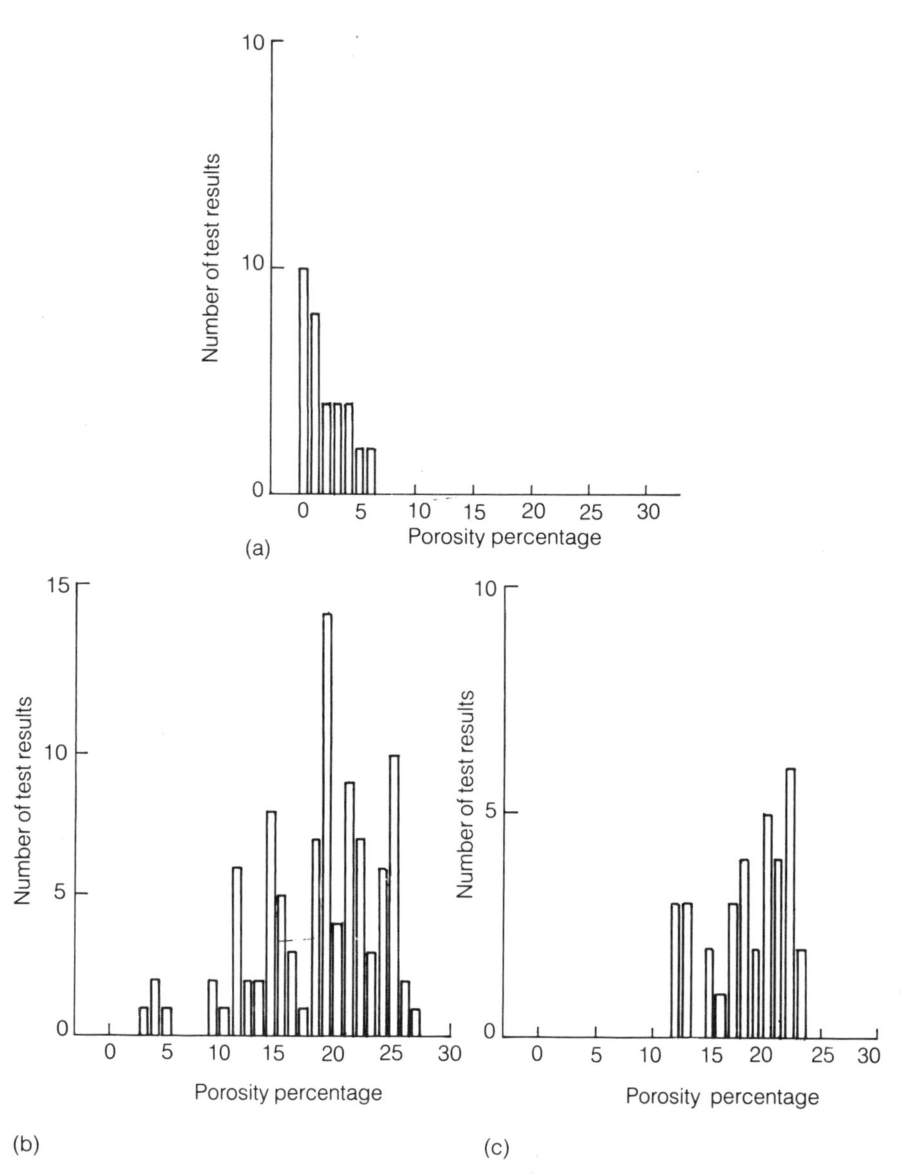

Figure 2.10 Porosity of limestones used for building: (a) Carboniferous limestone; (b) Middle Jurassic limestones: (c) Portland stone.

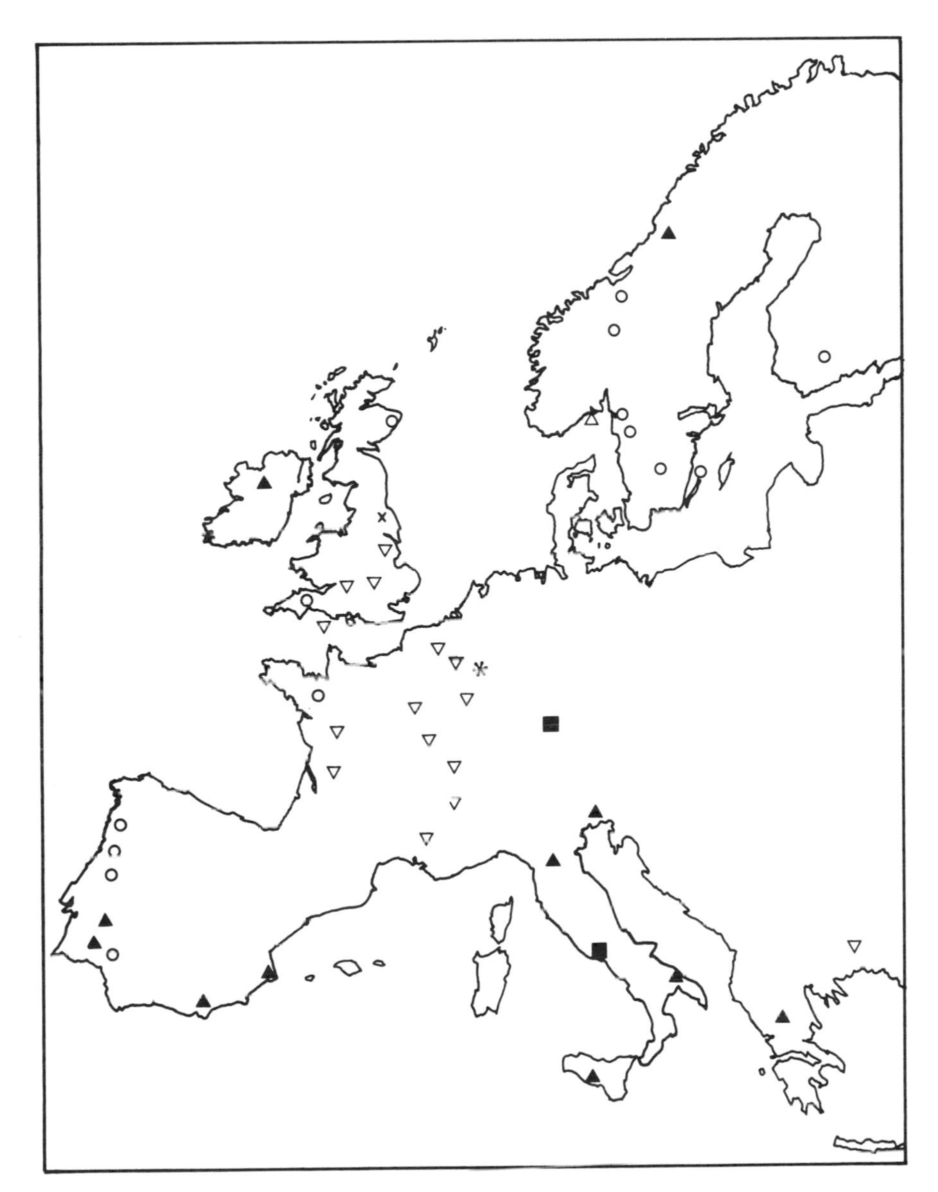

Figure 2.11 Map of western Europe showing the chief areas producing, or formerly producing, dimension stone. ○, Granite–gneiss; △, Syenite; ▽, Limestone; ▲, Marble; ■, Travertine; *, Volcanic rocks; X, Sandstone.

ornamental stones, including long-famous dark red and brecciated varieties.

Throughout the Alpine fold-belt of Greece and Turkey small occurrences of marble are widespread; often worked on a small scale, they include a very wide variety of colours and textures. One such deposit, that of the significantly named island of Marmora, can probably claim greater antiquity for its production than those of Italy. In contrast, although limestone is abundant in the Eastern Alps of Switzerland and Austria, there is no significant marble production – a reflection of the greater fracturing of these rocks which has resulted from the more intense orogeny of the northern region.

Spain and Portugal, on the other hand, with a geological history not greatly dissimilar to that of the other Mediterranean countries, have extensive marble production in Beja and Almeria.

An unusual calcareous rock, in which Italy is again the world's leading producer, is travertine. This is a deposit formed by hot springs, and its convoluted laminations reflect evaporation layers; its occurrence is limited to the actively geothermal areas of Lazio and Tuscany.

A comparison of the Mediterranean situation, and that of north-west Europe, is instructive. Belgium, for instance, has no true, metamorphic marbles; its extensive 'marble' industry is based upon hard limestones of Devonian and Carboniferous age. The Devonian rocks are principally reef-limestones of Frasnian age which occur as isolated mounds in a belt across the central Ardennes. They are largely red in colour, and owe their decorative variety to cross-sections of the reef-forming organisms (stromatoporoids and corals), to variously laminar sedimentary structures, and frequent calcite-filled geodes. The Carboniferous rocks, by contrast, are bedded, highly crinoidal limestones – the calcite crystals of the broken crinoids giving the rock a granular appearance which has been the origin of its trade name of 'petit granit'. Particularly famous in this category are a number of crinoidal limestones in which the fine-grained matrix is almost black – the 'marbres noires' of Dinant and Tournai, which were worked by the Romans, and exported all over Europe by the thirteenth century. The Dinant stone, in particular, readily splits into large, thin slabs – a consequence of its closely spaced bedding and widely separated joints – which has made it a great favourite for the construction of the tops of large sarcophagi.

In the British Isles, the only production of true marble lies in the west of Ireland, where some of the metamorphosed limestones of the

Dalradian sequence are of sufficient extent to support small-scale commercial extraction. The equivalents of the Belgian Devonian limestones are to be found in Devonshire, and the Torquay and Plymouth limestones have been extensively exploited, although workable reserves now appear to be largely exhausted or sterilized. A local equivalent of the 'petit granit', known as Hoptonwood stone, and quarried from the Carboniferous of Derbyshire, enjoyed a vogue in Victorian England; but although the very extensive development of crinoidal limestones in the Carboniferous of the British Isles contains many such ornamental stones, they have never been properly exploited.

It is at higher stratigraphical levels, in the Middle and Upper Jurassic, that the stone industry of Britain has had its major development. A variety of thick limestone formations occurs over almost the whole Jurassic outcrop from Humberside to the Dorset coast. The oolitic texture of many of these limestones allows them to be split with equal ease in any direction – a property which distinguishes them as 'freestone'. They have been subjected to only limited tectonism, their only structural feature being a widely spaced rectangular jointing, which is helpful in the extraction of large blocks. With a few exceptions, however, they will not take a high polish; so that their main application is in the form of hewn blocks and slabs for structural masonry or for cladding. The Middle Jurassic 'freestones' were quarried principally in the areas around Bath, north of Oxford, and in Lincolnshire; and they were a major constructional material in those areas relatively close to their sources. The Upper Jurassic Portland stone, by contrast, because its quarries had direct access to the sea, was carried longer distances, making a major contribution to the rebuilding of London in the seventeenth and eighteenth centuries. The Portland stone, unlike the Middle Jurassic rocks, is not strongly öolitic, but is a shelly limestone with a dense micritic matrix. It is possible that this dense matrix contributes to the resistance of Portland stone to acid city atmospheres, which is somewhat better than other, equally porous (Figure 2.9) limestone. One variety of Portland stone, known as Roach, contains numerous hollow fossil shells, partially filled by large calcite crystals; when cut this bears at least a superficial resemblance to the Italian travertine.

Purbeck marble, Sussex marble, Bethersden marble are curiosities – these are thin freshwater limestones found in Purbeck and Wealden beds of southern England, which owe their ornamental character to the closely packed gastropod shells of which they mainly consist. If the region had not been otherwise devoid of polishable stones, they would probably never have figured as extensively as they do in the church and domestic monuments of southern England.

The dimension stone production of Scandinavia is an example of an industry based on intensely metamorphosed and igneous rocks. The Pre-Cambrian complex of which most of Sweden and Finland, and some parts of Norway, are composed, contains large areas of gneissose rocks. Although these rocks display an alignment of their crystals – the foliation – their structure is so firmly welded that this does not give planes of weakness to the rock. The complex is also invaded by a series of large granite plutons, whose coarsely crystalline rocks, lacking any clearly defined linearity in their structure, also form excellent dimension stone.

These rocks, containing as they do a large proportion of quartz, are extremely hard to cut and polish, hard abrasives and even diamond being used in the process. By the same token they are extremely durable, and take a high polish. In this industry, transport costs of this very bulky material always outweigh processing costs, and on this basis Sweden is able to sustain an active domestic sale, together with export to stone-deficient countries such as Denmark, and to many other parts of the world.

Norway presents an interesting contrast, since a major part of the country consists of the Caledonian orogenic belt, formed in the Palaeozoic, and involving rocks in which fine-grained sediments predominate; in their metamorphosed state these produce rocks with well-developed schistosity and cleavage. Norway's production of dimension stone is thus confined to a few late-stage granitic intrusions. It does, however, have one unique rock – larvikite. This is an intrusive syenite body, consisting very largely of large crystals of the feldspar labradorite. This mineral contains abundant reflective cleavage planes, which produce an iridescent play of light within the rock – an extremely attractive effect that ensures that 'Norwegian Pearl' figures prominently on shop-fronts throughout western Europe.

Western Germany is an example of dimension stone production based largely on recent volcanic rocks. Volcanic rocks, particularly basic ones, do not normally figure largely in a dimension stone industry; they tend to be dark in colour, rather fine grained in texture, and their susceptibility to weathering makes quarrying of large blocks difficult. Nevertheless, coarse gabbroic rocks have been quarried extensively in the Rhineland region, and their dark colour gives a sombre aspect to major buildings – for instance, to the cathedral at Cologne.

In the Eifel region of West Germany are two unusual developments. The Eifel is a region of very recent explosive vulcanicity, and there are extensive basaltic and felsic lava flows. One of the latter is a pumaceous selbergite. This rock is cut and sawn with ease, and its vesicular nature makes it conspicuously light, so that it is used for lintels, door-jambs, sills, etc. in the region.

A number of the basalt flows have very well-developed hexagonal jointing; the combination of this with a well-spaced horizontal flow-jointing allows the rock readily to be split into hexagonal blocks 10–15 cm in thickness. These have been extensively used in the nearby Netherlands for a mosaic forming the surface of dyke sides, embankments and sea-defences. But basalt is an unsatisfactory rock in many ways, since its ferromagnesian minerals weather rapidly, forming swelling clay minerals, which cause the rapid disintegration of the rock; and this has been a subject of much concern recently in the Netherlands – which at the moment is importing some two million tonnes of basalt annually, not only from Germany, but from as far away as Czechoslovakia. Interestingly, so established has this hexagonal pattern become, that concrete blocks are now manufactured in this shape for these same purposes.

As an example of a dimension stone industry based on sedimentary sandstones, we can look to the north of England. The largest concentration of construction stone quarries in the British Isles is in the Pennine region of West Yorkshire. Here, medium to fine-grained sandstones occur at intervals throughout the Namurian and coal measure strata. Unlike many sandstones, they have little calcareous cement, and so resist solution in polluted atmospheres – an important feature in their extensive use in the industrial towns adjacent to their occurrence. The absence of clay minerals, and of calcareous cement, gives them compressive strengths of around 60 MPa (higher than most other sedimentary rocks), while porosities, at an average of 10%, are low (Figure 2.12). The occurrence of parallel bedding planes at 2–3 cm spacings allows much of the rock to be split into flagstones; as paving in cities throughout Britain they dominated the market in the early part of this century, and have still not wholly been replaced by the concrete slab. The presence of mica on the bedding surfaces, which is the principal cause of the splitting, also gives an attractive sparkle; and the angularity of the sand grains contributes a non-slip factor which is not unwelcome in the British climate.

In contrast, the younger Permo–Triassic sandstones of the north of England, quarried as Penrith and St Bees sandstones, and equivalent strata in other parts of England and Scotland, have lower

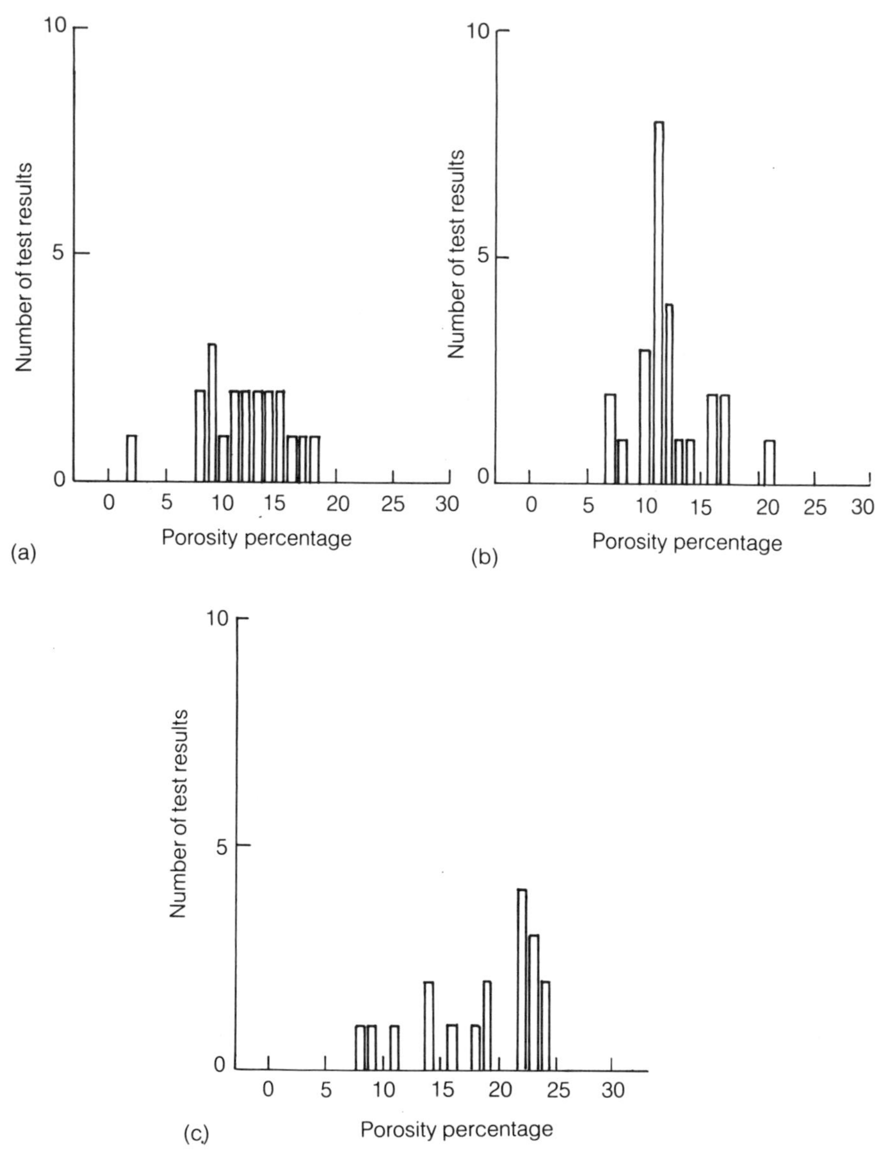

Figure 2.12 Porosities of British sandstones used for building: (a) Carboniferous (Millstone Grit); (b) Carboniferous (coal measures); (c) Permo–Triassic.

strengths (Figure 3.1) and higher porosities (Figure 2.12); and many older buildings in which these have been used now display extensive erosion.

The dimension stone industry is thus seen as one in which costs – of extraction, of processing, and of transport – are always high; and one in which automation offers little prospect of reducing costs. It follows that only deposits which are geologically excellent – ones which are extensive, uniform and where structures assist in the extraction – are likely to be worked. Compared with crushed aggregate production, it affords a low profitability, and there is no doubt that many rocks suitable for dimension stone production are today being fed into the jaw-crushers of aggregate plants. The best hope for the industry appears to lie in the tandem development of dimension stone and aggregate, so that the same quarry can use its best material for cutting into blocks, and the waste from this process, together with the lesser-quality material, can go to the aggregate plant.

2.5 SLATE

The facility of a rock to be split into large, thin sheets suitable for roofing is in general found only where slaty cleavage is developed. Cleavage is due to the realignment of platy minerals in response to pressure; structural geologists recognize many kinds of cleavage, but only the large-scale regionally developed slaty cleavage is responsible for giving rise to workable slate deposits.

The original sedimentary rock must have been almost entirely argillaceous; regional pressure, accompanied by some rise in temperature, converts the clay minerals to chlorites and micas, which orientate themselves with their platy direction at right angles to the principal stress. Any quartz particles present will not be affected by this process, so the presence of any amount of silt or sand in the rock prevents the proper development of the cleavage planes. Thick accumulations of mudstones of this kind commonly occur as distal turbidites, so that they have a laminated structure, with individual laminae showing slight increases of silt or sand content. Since the angle of cleavage is almost always different from that of the bedding, these laminations show as an attractive striping of the cleavage surface. If a bedding lamina contains too high a quartz content, it will form a limit to the cleavage development, and thus to the size of the slate. In practice, blocks of slate are split out from the rock-face along such bedding planes.

The geological environment in which such rocks were deposited must be that of deep oceanic water; in this situation of anaerobic conditions the presence of sulphides, and particularly that of iron pyrite, is to be expected. Under metamorphism, this will recrystallize to large cubic crystals, or aggregates of crystals, which are often a prominent feature on the slate surface. Their presence is deleterious, since pyrite rapidly weathers on exposure, leaving a weak spot, or even a hole, in the slate. Fortunately, not all beds of mudstone contain high quantities of pyrite, so that the producer is able to select the more favourable seams for production.

The other geological requirement is that other structural features imposed on the rock must be at a minimum; more intense metamorphism will produce other cleavages, fractures and closely spaced jointing. In the best slate rocks, there is only a relatively widely spaced jointing, roughly at right angles to the bedding and/or the cleavage, which with the bedding provides the limits of the extracted block, and thus determines the size of the slate.

Extraction of a block of slate from the quarry face is carried out as with any other dimension stone; but the splitting of this block into slates is still performed entirely by hand, as no mechanical method has yet been devised. After splitting, the slates are sawn to size.

It will be evident from the above that this an industry which produces very large quantities of waste; in fact a better-than 40% useful rock production is rarely achieved. The large waste-heaps which disfigure the landscape in most former slate-producing areas are very visible evidence of this. Local craft production of cuckoo-clocks and ashtrays can make little inroads into these accumulations; and most slate-producing areas have sought actively to diversify their markets.

Blocks not sufficiently fissile to split into roofing slates can be slabbed and polished, and used for cladding and floor tiles, hearths and lintels. There appears to be yet no substitute for slate as a bed for billiard tables, and Italian producers, whose slate is less strongly metamorphosed and therefore more readily cut and polished, have found a useful market here. Roughly crushed, the waste can be used as rock-fill (section 2.8) and as a road sub-base; but its comparative softness and ease of polishing preclude its use as a road-surfacing material. Ground to <2 mm size granules, it is sold as a surfacing material for roofing felt. Ground to a fine powder, it is marketed as a filler for plastic and rubber; ironically, this powder forms the base of a product which may become a serious competitor for roofing slate – incorporated into a resin base, and moulded against a natural slate, it produces a realistic simulation to the natural product at a competitive

Figure 2.13 Slate-producing areas of north-west Europe: 1, Lake District, England; 2, North Wales; 3, North Cornwall, England; 4, Central Brittany, France; 5, South-west Ardennes, Belgium; 6, Eifel, Germany; 7, Valongo, Portugal; 8, Orense, Spain; 9, Leon, Spain; 10, Segovia, Spain; 11, Triora, Italy.

price. A further use for powdered slate is in the manufacture of lightweight aggregate (section 5.9).

From the above it will be evident that the geological conditions to provide a good roofing slate are precisely delimited; and while 'slate' in the geological sense is a common rock, workable deposits are not. The United Kingdom has three such areas – North Wales, the Lake District and Cornwall (Figure 2.13).

In North Wales Cambrian rocks of the Snowdon region, and Ordovician rocks around the Harlech Dome provide the most significant production. In the Lake District there are black Silurian slates, but the most characteristic products of the area are the green Westmoreland slates which have their origin in the Borrodaile Volcanic Series, of Ordovician age. The majority of this series consists of lavas and tuffs, but at two levels there occur fine-grained ashes which have taken on a good slaty cleavage.

In Cornwall the Delabole slates of the Wadebridge–Camelford area are of Devonian age. This is a revealing example of the dependence of slate of commercial use upon the precise metamorphic state of the rock. 'Slates' of identical age occur both to the north and to the south in this Upper Palaeozoic orogenic belt. The equivalents to the north, the Pilton beds, have been only shallowly buried, and have assumed only a rough 'pencil cleavage'; while those to the south, included in the Mylor slates, have been strongly tectonized, and contain several succesive cleavage and fracture planes which render them commercially useless. The deep weathering which has affected all the rocks of south-west England has permitted the percolation of iron-rich waters down the cleavage and joint-planes of this formation, giving the surface of the slate an attractive range of red and brown coloration. This feature is effectively used for indoor decorative stone, particularly for hearths and fireplaces.

In Britain, as in many parts of Europe, there has in the past been much use of other locally available fissile rocks as 'slate' for roofing. These are usually thinly bedded limestones (as in the 'Stonesfield slate' of Oxfordshire) or sandstones (the Horsham stone of Sussex); their fissility derives from their bedding laminations, not from any slaty cleavage. Such 'slates' are generally very variable in thickness, and while contributing much to the picturesque scene, are now of no commercial significance.

In Western Europe slate production is almost entirely confined to the Palaeozoic massifs; a single, well-defined slaty cleavage is not found in either the Pre-Cambrian or the Mesozoic orogenic belts. (The only exception to this latter is the slate production at Triora, in the Maritime Alps of Italy.) In Belgium, present-day slate production is confined to the Lower Devonian slates of the south-west Ardennes, while the same formation is exploited in the adjacent Eifel region of Germany. In Central Brittany the rocks are highly tectonized ?Lower Carboniferous. In all these developments within the Variscan fold-belt (including Cornwall, Brittany, the Ardennes, the Eifel and the former producing areas of the Harz Mountains, etc.) bedding and jointing are closely spaced, so that only small-sized slates are produced – a feature which gives a distinctive character to slate roofs over the whole of Central Europe. In the Iberian Peninsula, Ordovician slates are exploited in north Portugal and the adjacent Orense region of Spain, while other Spanish producers are in Leon and Segovia.

2.6 ARMOURSTONE

A major use for stone in large sizes is as armourstone, in the construction of sea-defences and breakwaters. Such constructions use large quantities of stone, often in very large sizes (blocks up to 20 tonnes in weight are used), so that proximity of site to rock source is imperative. The strength and durability of natural rock is an important asset in this context, and must be taken into consideration when its use is considered against its major rival, concrete (Chapter 6). Although steel and timber have been used for this purpose, 95% of breakwaters in the world use stone or concrete, or a combination of the two.

Most rock breakwaters are built of a central core and an outer armour. The core is composed of 'run of quarry' stone dumped on the sea-bed, generally up to sea-level, so that it can be built outwards from the shore. This core mainly consists of cobble- or boulder-sized material (i.e. between 60 and 500 mm in diameter). On this the armour layers are laid. Capping the core is usually a filter layer, composed of gravel-sized rock (2–60 mm in diameter) but often containing also coarse sand. The object of this filter layer is to prevent fines from being washed out of the core, which would cause settlement and weaken the structure. On top of this are one or more layers of rock fragments, often a secondary layer below, composed of relatively small fragments, on top of which a primary layer, composed of the largest blocks, is constructed. Alternatively the primary layer may be constructed in concrete, or in masonry.

The roughly crushed rock for the core, as produced by the quarry, may contain a variable amount of fines (i.e. of coarse sand or smaller size), and these will largely be washed away during the emplacement process. The amount of fines depends upon the lithology of the original rock; in general hard igneous rocks and sandstones produce little fine material on crushing, whereas limestones tend to produce large quantities. This is a significant factor in the many areas of the world where limestone is the only available rock.

The design of the filter layers is directed towards achieving a particle-size distribution which gives the minimum of voids, and this can be calculated from well-established formulae. However, the shape of the particles is also important, although much more difficult to measure, and it can be shown that flaky or elongate particles should be avoided. Rocks which produce such particles are those which have schistosity or lineation, and this factor needs to be taken into account in the selection of a rock source for the filter layers.

The size of the blocks used in the primary armour is important in determining the effectiveness of the breakwater, and of the slope at which it can be maintained. Initially, this is determined by the fracture spacing of the rock *in situ*, which can be ascertained by visual inspection of the quarry, or by observations on cored boreholes (Rock Quality Designation, RQD), or, to a limited extent, by geophysical measurements. The actual size produced, however, is also controlled by the methods of extraction, and by any primary crushing which is undertaken.

It will be obvious that use as armourstone places the greatest demands on a natural rock, which must have a great measure of resistance to exposure in this very aggressive environment. These qualities may be defined as toughness, i.e. resistance to fracture under impact; hardness, i.e. resistance to abrasion; and durability.

The primary agents attacking the rock, as with dimension stone uses (section 2.1), are those of physical and chemical weathering, but these tend to be aggravated in a breakwater by the presence of salt, and by the frequent changes of wetting and drying consequent upon the rise and fall of the tides. In addition, with the physical impact of waves in times of storm, even the largest blocks are liable to be moved; while cobble- and smaller-sized fragments may be in constant motion. It has long been recognized by sedimentologists that this zone represents the environment where the greatest degree of abrasion and attrition takes place.

In a coastal marine environment four vertically arranged zones can be recognized (Figure 2.14). At the base (zone IV) is the permanently submerged zone, below low-water mark, where the armour is protected from subaerial weathering agents, although still exposed to strong water currents generated by the breaking waves. Above (zone III) is the intertidal zone; here daily saturation and drying provide the mechanism for accelerated physical (from crystallization

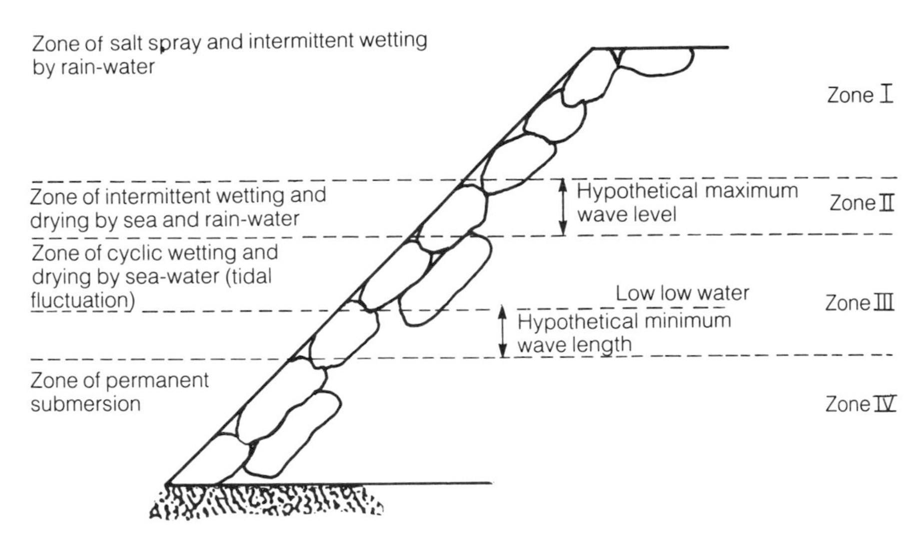

Figure 2.14 The four main weathering zones of the coastal marine environment (after Fookes and Poole (1981, Figure 1)).

of salt crystals) and chemical corrosion. It is also the zone in which the breaking waves exert their strongest force to move, and therefore abrade, the armourstone. Zone II is above normal high-water mark, but still subject to alternate wetting and drying by waves, and to abrasion by wind and wave-thrown pebbles and sand. In zone I, the presence of salt spray, and continuous exposure to subaerial forces, result in accelerated weathering action. In all zones, biological agencies may assist the erosion – boring and excavating animals in zones III and IV, seaweeds and land plants at other levels.

The three qualities of hardness, toughness and durability may to some extent be measured by laboratory tests, although, as with dimension stone, the transfer of the laboratory results to actual performance is as yet imperfect. The unconfined compressive strength is clearly important as a measure of the ability of the rock to withstand repeated impact; and in this environment it is necessary to measure the sample both in the dry state, and when saturated by fresh- and sea-water. The wet and dry strengths are often significantly different, and the ratio of wet/dry strength has proved of value in the assessment of rock-suitability. A point-load test, which measures the impression made on the rock surface by a steel point under constant load, is also potentially useful.

There are many tests available to quantify the resistance of a rock to abrasion, but almost all are devised for use in the aggregate

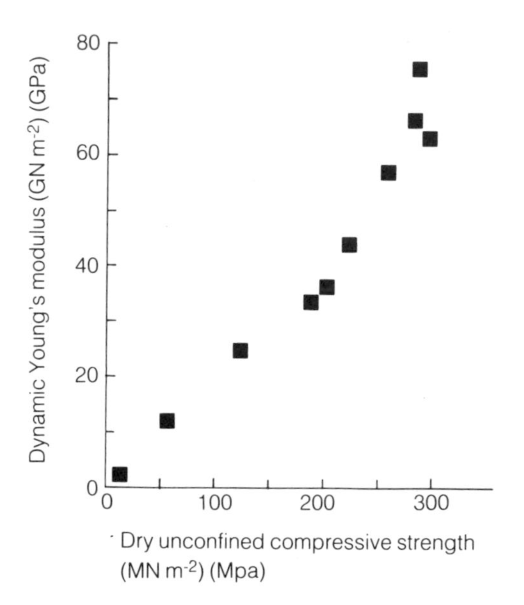

Figure 2.15 Plot of Young's modulus against compressive strength of Dartmoor granite samples of various weathering grades (after Dibb *et al.* (1983, Figure 13)).

industry, and involve measurements upon crushed fragments. It has been suggested that in large blocks such as those used on breakwaters, the pattern of fractures is the most significant feature, and tests using the modulus of elasticity, combined with that of unconfined compressive strength, have been devised (Figure 2.15). In practical use, the simple measurement of specific gravity appears to relate closely to abrasion resistance. Other tests which use the loss of weight or particle size by autogenous milling under fixed conditions have been shown to be practicable, and since they simulate 'real' conditions, ought to be an accurate measure of resistance to abrasion.

It will be clear to the geologist that the resistance of a rock to abrasion must depend upon its mineralogy and its texture. The mineral quartz, by a simple scratch test, can be shown to be harder than calcite, so that limestones should be easier to abrade than siliceous rocks. A fine-textured rock with well-developed schistosity should be abraded more easily than a coarse equigranular one. A sandstone with a pressure-solution cementation should resist abrasion more readily than a loosely, calcite-cemented one. Thus detailed petrography is one of the most important assessment methods in the prediction of suitability for armourstone.

Tests of durability, as with dimension stone, are not easy to relate to practical experience. The consensus is that the main factors in the natural weathering process are the mechanical breakdown caused by crystal growth in the pores of the rock, and the chemical conversion of aluminosilicate minerals to clay. The first process ought to be related to the porousness of the rock, which can be estimated from its specific gravity, or from a variety of tests involving total or partial saturation by water (section 2.1). These tests, however, measure the total volume of pores in the rock, and there is evidence to suggest that the size, and particularly the maximum size, of the pores may be the more important factor. Tests involving cycles of freezing and thawing, or wetting and drying, and subsequent measurement of weight loss, are widely used, on the grounds that they are practical simulations of real life.

In practical terms, one of the most important features in the selection of rock for armourstone is the recognition of the effects of weathering of the *in situ* rock. Weathering effects can reduce the compressive strength of a rock from exceeding 250 MPa to values of almost nil (Figure 2.15). Standard practice in engineering geology recognizes seven progressive grades of weathering; the higher grades are easily recognized, but slightly weathered and moderately weathered rocks can be less easy to identify, and yet these grades may have significantly lower strengths, and careful observation of the quarry face or of borehole cores is essential. Since the decline in strength is related to the development of clay minerals, and particularly of expanding clay minerals, laboratory measurement of these gives an objective measure; this can be done by X-ray diffraction, or, more simply, by staining with methylene blue. The weathering problem is particularly acute in basic igneous rocks, and the Netherlands, where imported basalt is used very extensively, have developed tests in which the specimen is milled and then tested for clay minerals, thus measuring its actual and potential weathering loss (section 2.1).

2.7 ROCK FOR FILL

Fill is a general term to describe granular material used in an unbound state to create a foundation, whether for a building, a road, or a runway. Some artificial materials are used in this way: crushed concrete, burnt colliery shale, and furnace slag; but natural rock is still the major source, and as such represents the largest volume of quarry output for any purpose.

In road construction loose fill is used to provide a capping layer and a sub base. In the construction process the soil and any loose material is removed to form a level surface (subgrade or formation level). If this surface is not firm a capping layer, generally of 'run-of-quarry' or roughly crushed stone is laid on it; on top of this is laid the sub-base composed of selected fill. The roadbase, basecourse, and wearing course, which follow in sequence are composed of aggregate bound with cement or bitumen (Chapter 5).

The object of the sub-base is to spread the stress created by the traffic over a wide area, so that the subgrade below is not deformed. The selected fill used must therefore be able to withstand the constant and the intermittent load imposed by the passage of vehicles; it must also be able to sustain the interparticle movement generated within it. British Department of Transport specifications recognize two types of material for sub-base, defined by particle size and content. Type 1, specified for heavily trafficked roads (defined in this instance as more than 6 million standard axles per year), excludes natural sands and gravels, which are allowed in Type 2 – this also allows higher percentages of sand in the grading.

When the road is finished, the upper layers provide some protection for the sub-base; nevertheless water can freely pass through the open structure, so the constituent rock must be able to maintain its properties when in the wet state. It must also be able to resist the weathering processes, both mechanical and chemical. During the construction process the sub-base is used as the track along which heavy construction machinery can run; it must therefore be able to withstand the weight and impact of such vehicles. During the construction phase it may remain for a long period exposed to the elements, so that its weathering resistance is even more important.

The grading specification (Figure 2.16) is designed to create a platform which is rigid enough to bear the imposed load, but at the same time to provide free drainage of water through the sub-base. It is particularly important to avoid the inclusion of clay, whose presence

1. allows the sub-base to deform
2. clogs the voids in the structure, and
3. in extreme cases can be forced upwards into the roadbase and basecourse, causing their disruption.

Despite the manifest importance that the components of fill should be hard, tough and durable, there is little standard practice in the testing of these parameters. Methods which are applied to dimension stone, and which measure intact rock (unconfined compressive strength, porosity etc.) and methods used for testing aggregates (AAV, AIV, etc.) (pp. 72–4) are used, and are subject to the same

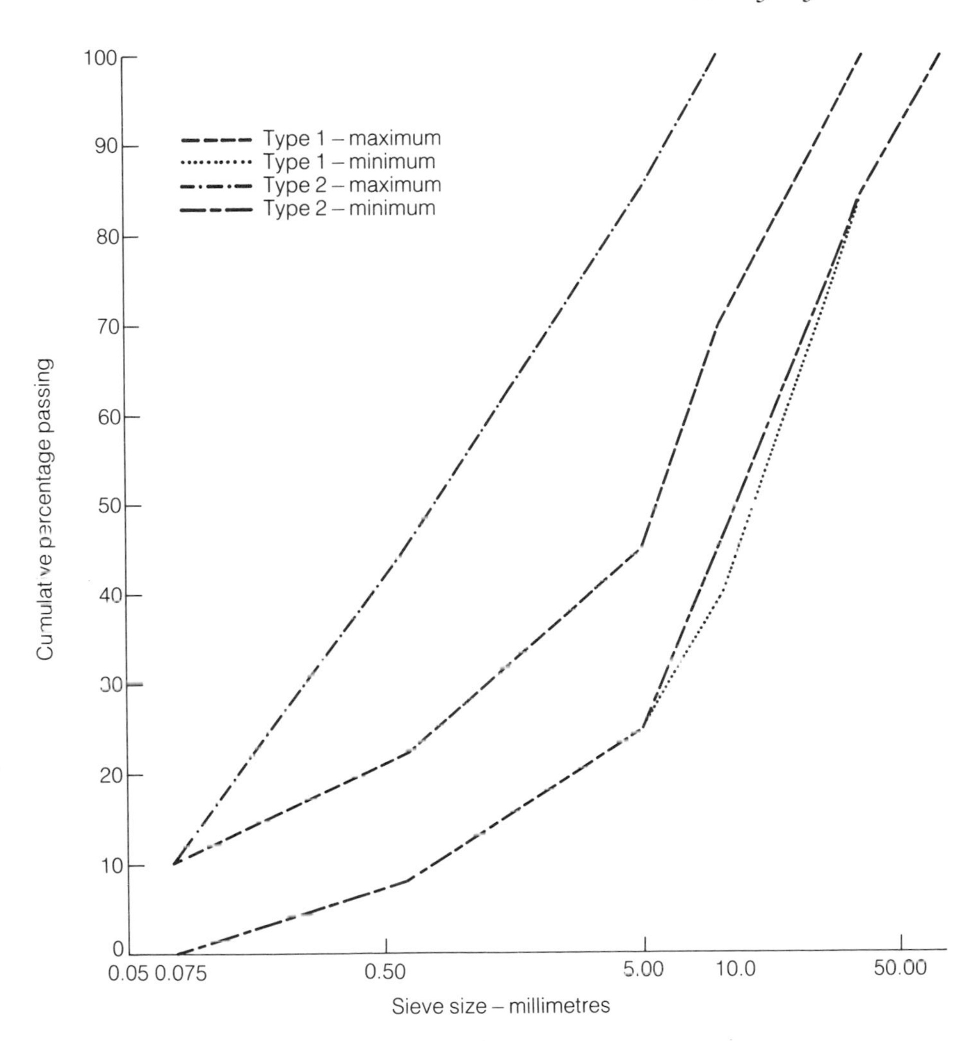

Figure 2.16 Grading limits of specification for granular sub-base. From Department of Transport *Specification for Road and Bridge Works*, 1976, Table 8/1.

reservations detailed. Probably the most severe disadvantage in the application of all these methods is that they are normally applied to dry, or partially dry rock; whereas a fill material must be able to maintain its strength even when repeatedly saturated. It has been shown that the loss in strength on wetting, measured by Aggregate Crushing Value, can range from 20% in hard gravels to 80% for some limestones; measured by Aggregate Impact Value from 17% to

35%, and by the 10% fines test from 24% to 63%. These figures are so large that clearly any application of dry strength measurements has little meaning in the context of fill.

A test addressed to this problem, but particularly in mudrocks, is the Slake Durability (SD) Index. In this fragments of the rock are rotated in a wire-mesh drum under water, and the weight loss measured. This weight loss, reported as a percentage of the initial weight, is the SD index. The Washington Degradation Test arrives at a Degradation Index in a somewhat similar manner.

Because the fill may be exposed for some length of time during construction, and in any case may be subjected to frost action, the frost susceptibility is regarded as important. This can be measured directly by refrigeration.

A large number of failures in road sub-bases have been traced to the use of partially weathered rock. A rock may, on visual inspection, appear sound and unweathered, but alteration of the primary minerals to weaker, secondary ones may substantially reduce its strength in use. In granites and similar rocks, the decay of feldspars to kaolinite is an important cause, in basic igneous rocks alteration of ferromagnesian minerals to serpentine and smectitic clay. While perhaps most significant in igneous rocks, similar changes can take place, for instance, in arkosic sandstones.

Fortunately, these changes are readily detected under the petrographic microscope. Using a point-count technique, it is possible to devise a **micropetrographical index** which compares the number of sound with unsound mineral grains. Tested on a granite, the method shows good correlation with other engineering test methods.

Different climatic conditions can affect the type and speed of weathering, and it has been found necessary to take this into account in determining the suitability of a rock for use in a particular area. A particularly dramatic weathering effect, which has affected housing foundations and highways, occurs when pyritic shales are used as a foundation material. Pyrite (Fe_2S_x) readily oxidizes to sulphur trioxide, which combines with water and calcium carbonate to form gypsum. The crystallization of large selenite crystals produces a volume expansion, with disruptive effects on roadbase or concrete screed laid on top. Identification of pyrite in thin or polished section, or even hand-specimen, is normally easy; and it can be readily recognized in the field by the occurrence on weathered surfaces of the sulphur-yellow mineral jaroszite. In the laboratory it can be quantified by a heavy-mineral separation of the ground shale, or by a chemical analysis of the sulphur content. Tolerance limits are unknown, but it seems likely that any shale containing more that 1% or 2% pyrite should not be used for fill.

2.8 ROLE OF THE GEOLOGIST

The foregoing discussion has shown that the qualities which determine the suitability of a rock for use as a building stone, as an armourstone, or as a constituent of rock-fill may be directly related to its mode of formation and geological history. Armed with this knowledge the geologist can make an informed search for new deposits, and can direct quarry operations to the maximum benefit.

It can also be seen that there is a clear, but not simple, relationship between the petrographic characters of a rock, and its performance as part of a building or engineering structure. It is significant that good correlation between the various engineering test methods, and correlation with petrographic characters, is only achieved when the rocks studied are of comparable lithology. Given the great variety of mineralogy and texture of rocks, it is unreasonable to expect that they would respond in the same way to external forces. Thus the effect of impact upon, say, a glassy rhyolite, must be entirely different from the effect on a crystalline granite, or a calcite-cemented sandstone – so to expect this to be measured by a single coefficient, which can then be compared with other characters such as hardness, is not logical. However, within any one lithological type there is consistency of test results, and test methods can be used to predict its behaviour.

In my view, in assessing the potential of a rock for use, the first requirement is a full petrographic study of the rock, identifying its mineralogy, grain size, texture, fabric and the weathering state. Only then is it possible to decide the ways in which that rock might fail, and thus decide which tests are appropriate to determine its suitability.

3

Coarse aggregate

3.1 INTRODUCTION

Because they have different functions, and therefore different end-products, it is convenient to follow industrial usage in separating coarse aggregate (that is to say, rock particles over 5 mm in diameter) from fine – although, of course, their geological occurrence is often the same. Such sizes of particles may occur naturally, as in glacial and alluvial gravels, and therefore only require separating; or they may be created by rock crushing.

Coarse aggregate is mainly used as a primary ingredient in concrete, or as a constituent of roads, either in the body of the road or as surface chippings. In most countries of the world, it represents the greatest bulk, and often the greatest value, of any mineral extracted. Present-day construction is based on the availability of coarse aggregate at a low price; and transport of this heavy, bulky material is a major element of the cost. Sources of such aggregate thus need to be sought close to their point of usage.

3.2 AGGREGATE PROPERTIES AND TESTING

In the various ways in which aggregate is used, it is exposed to a variety of stresses; and the properties of the road, bridge or building in which it is used will largely depend upon the properties of the aggregate. It needs to react favourably with the cement or bitumen in which it is embedded; it needs to resist heavy loads, high impacts and severe abrasion; it needs to be durable in the prevailing environmental conditions. And these properties will need to be tested and assured before the road is built or the concrete is poured.

For this reason, engineers have developed an array of tests designed to predict the behaviour of the aggregate; these tests are by nature empirical, and often are simulations in the laboratory of working conditions. This pragmatic approach tells us little of the fundamental reasons for rock properties; and while in theory it ought to be possible for detailed petrographic study of a rock to predict its behaviour, in practice this is not yet possible.

The properties to be tested are many and various; they may be subsumed under the headings of strength, water absorption, shrinkage, abrasion resistance, abrasiveness, polishing behaviour, flakiness and resistance to weathering.

Strength in any material may be described as its ability to resist compression, shear stress, or tension; and in homogeneous materials there is a reasonably precise relationship between the three. In rocks, which are almost without exception inhomogeneous, this relationship is more complex, and for this reason, as well as the great problems involved in its measurement, **tensile strength** has received little attention in the construction industry. **Compressive strength**, however, can be measured quite simply by applying pressure in one direction to a cube of rock, and recording the pressure at failure. This is **uniaxial** or **unconfined compressive strength** (UCS).

The compressive strength of a rock ought logically to be dependent upon

1. the strength of the individual minerals, and
2. the strength of the cementing materials of the individual grains

while it might be expected that mineral strength might in turn have a relationship with mineral hardness, as measured by the familiar Moh's scale. These suppositions are to some extent borne out by the data recorded in Figure 3.1, which shows the range and frequency of measurements of compressive strength made on a variety of British rocks.

The weakest rocks used for aggregate in the British Isles are the Jurassic limestones of southern and eastern England, being 'weak' to 'moderately strong' (5–60 MPa). They are composed entirely of calcite, a mineral whose three cleavages give multiple lines of weakness to the crystal structure; and the rocks consist primarily of shell or öolitic fragments bound together by a calcite cement. Only slightly stronger (20–80 MPa) are the Permo–Triassic sandstones of the Midlands and the North; although their primary mineral constituent, quartz, lacks cleavages and is thus inherently stronger than calcite, these sandstone often have a weak calcareous cement to the quartz grains. The Carboniferous sandstones, and Carboniferous limestones, have suffered much greater compression than their younger equivalents, and so their cement is dense and compact. A significant instance of the effect of the cement is seen in the Pennant sandstone (Figure 3.1, P), a Carboniferous sandstone from South

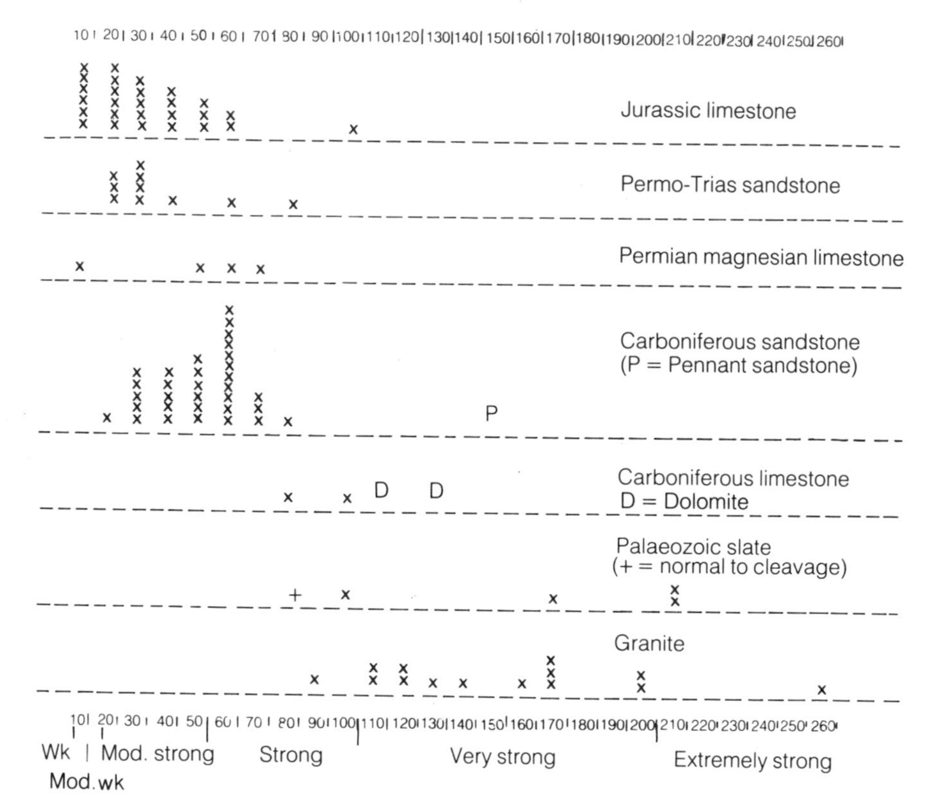

Figure 3.1 Unconfined compressive strength (UCS) of British rocks. Data from *National Stone Directory*, 6th edn, 1984 and other sources. Scale in MPa. Descriptive scale after McLean and Gribble (1979).

Wales, whose UCS of 160 MPa makes it one of the strongest of sedimentary rocks. This can be attributed to the fact that the sand grains are cemented by 'pressure solution', which produces a texture in which adjacent grains are keyed-in to one another, and any additional cement is entirely siliceous. The strength of Palaeozoic slate is interesting; a dense, compact rock in which silica and chlorite minerals predominate, it is very strong when measured parallel to the cleavage, but reduced to the middle of the strong range at right angles to the cleavage. The range of strength values shown by granite is from strong to extremely strong, and includes some of the strongest rocks known; the length of the range of values probably reflects the variety of rocks subsumed under this name, many of which are not granites in the strict petrographic sense. Those in the strongest categories are leucocratic granites, with high quartz content, good crystallinity, and a close, interlocking texture. Those at the low end of the scale are igneous rocks with higher feldspar and ferromagnesian mineral content.

A significant reduction of compressive strength is consequent upon the effects of weathering. Surface weathering effects, even in temperate climates, often extend many tens of metres into the rock; and even rock which is apparently fresh often contains weathering features which substantially reduce its strength. In granite, which has been extensively studied from this aspect, these features include microcracking of the quartz, and decay to kaolinite of the feldspar. Since these are readily recognizable under the petrographic microscope, it has been possible to develop a quantitative micropetrographic index, which can be used to measure the degree of weathering, and thus the weakening of the rock structure.

The apparatus for the direct measurement of compressive strength is expensive and cumbersome, and requires the cutting of a shaped piece of rock in the laboratory. In consequence a number of methods have been developed to measure this parameter indirectly, with apparatus which is simpler, and sometimes sufficiently portable to be used in the field. The Franklin point-load apparatus tests the breaking point of irregular rock pieces between two points; the cone penetrometer measures the degree of penetration of a steel cone under a standard load There is good evidence of correlation between the unconfined compressive strength of a rock and its modulus of elasticity (E), and rapid laboratory measurement of this E-modulus is now possible. This parameter can also be measured in the field, using the Schmidt rebound hammer.

It has to be recognized, however, that compressive strength is a measure of intact rock, and not of the likely performance of such a rock when broken into aggregate. It has thus fallen into disuse in the aggregate industry, being replaced by other tests more specifically related to rock fragments.

Most rocks which are used for road or concrete aggregate are quite resistant to chemical weathering processes; and the fact that these rock fragments are mostly embedded in concrete or coated with asphalt affords them a degree of protection. The basic igneous rocks, of which basalt is the most abundant, are, however, susceptible to weathering; in particular their ferromagnesian minerals are altered readily to swelling smectitic clays, resulting in a disintegration of the rock. Not all basalts are equally susceptible, and a test has been devised in which the formation of smectite is accelerated by fine milling, and then the amount of smectite formed in a standard time measured. Grinding under anhydrous conditions (in cyclohexane)

(Fig. 2.3) shows that a 'good' basalt will not generate smectite in the absence of water, but a more susceptible one will; study of these X-ray diffraction patterns shows too that the weathering process also involves the production of amorphous clays.

The possible reactions between aggregate fragments and their containing cement is a matter of great concern in the construction industry; this is discussed in detail in Chapter 6.

Water absorption and **shrinkage** are related phenomena. Different rock types show a differing capacity to absorb water. This is important in the production of concrete, since the absorption of water into the aggregate produces a shrinkage in the concrete, which in turn leads to cracking, and therefore unsoundness. A clear correlation can be shown between the drying shrinkage of concrete and the water absorption of the aggregate used; and a similar relationship appears to exist in bituminous mixes.

Rocks with a high absorption are basalts, dolerites, felsites and mudstones, while granites, limestones and flint show low absorption. While porosity might be expected to have a role in this relationship, in fact the presence of swelling clay-minerals seems to be the important factor. These are formed in basic igneous rocks from the decomposition of the ferromagnesian minerals, and it is significant that a weathered basalt will show a high absorption, while the same rock in the unweathered state will have a low value.

Abrasion resistance, **abrasiveness**, and **polishing** are properties of great importance in road-building. A rock which is going to be used to surface a road needs to withstand

1. the sudden application of intense pressure and stress – impact – by heavy vehicles
2. the grinding action of tyres – abrasion
3. the weathering processes – especially frost.

Before the days of heavy motor traffic, road surfaces were laid from natural aggregate; but the modern motor-road, with its continuous load of heavy motor vehicles, demands a much more durable material, and today crushed rock aggregate is usual. Since the performance of different rock types in this situation is so variable, engineers have developed a set of tests to measure a variety of parameters, which in turn are used to define standards for road applications. Methods and standards vary widely from country to country – in the following account only current British and American usage is described.

Aggregate crushing value (ACV) measures the percentage of fines (defined as material <2.36 mm) produced from a sample of aggregate by the application of a continuous load of 400 kN for 10 minutes. The related **ten per cent fines value** measures the load required to produce 10% fines from the aggregate sample. In both cases a lower value indicates a rock more resistant to crushing.

It would be anticipated that these values would show a correlation with the measures of compressive strength described above in this section; but this does not always seem to be the case, as a comparison of Figures 3.1 and 3.2 will show. Among sedimentary rocks, limestones (which in Figure 3.2 include both Carboniferous and Jurassic) have the worst ACVs, while the quartzites and gritstones, while not greatly different in UCS, are significantly better than limestones in ACV. Also the granites, which have the highest UCS values, are apparently less good in their ACV rating, and are outperformed by basalts. While it is not surprising that the 'Hornfels group' – which includes a number of very close-textured, siliceous rocks, with a dense interlocked structure, should perform best in the ACV test, the other results detailed above are somewhat anomalous, and suggest that the performance of a rock as an aggregate may not be wholly dependent on its compressive strength. It is known that the shape of the aggregate fragments known as flakiness (defined below) has an influence. Since the process of crushing used in the test methods, also causes the fragments to rub against one another, it is likely that an element of abrasion resistance (defined below) is also being measured.

The **aggregate impact value** (AIV) is a measure of the amount of fines produced by dropping a hammer of standard weight (13.5–14.1 kg) from a standard height (381 ± 6.5 mm) 15 times on to a sample of aggregate. The similarity of the results of this test to that of ACV (Figure 3.3) suggests that similar properties are being measured. The somewhat worse performance of quartzites and gritstones under impact, compared with compression, may be an effect of the numerous grain-boundaries in such rocks.

The **aggregate abrasion value** (AAV) is a measure of the loss of weight when a sample of aggregate, cemented to a fixed base, is abraded by a rotating lap fed by sand. The **Los Angeles abrasion value**, widely used in America, measures the weight loss during rotation of a loose sample in a steel cylinder containing steel balls. There is a demonstrable correlation between abrasion and compressive strength, and between abrasion and ACV (Figure 3.4), showing that to some extent similar properties are being measured. A similar ranking of the various rock types (Figure 3.5) is achieved by AAV as for other values. However, it is to be noticed that quartzites (a category including rocks with a mainly siliceous cement) are notably

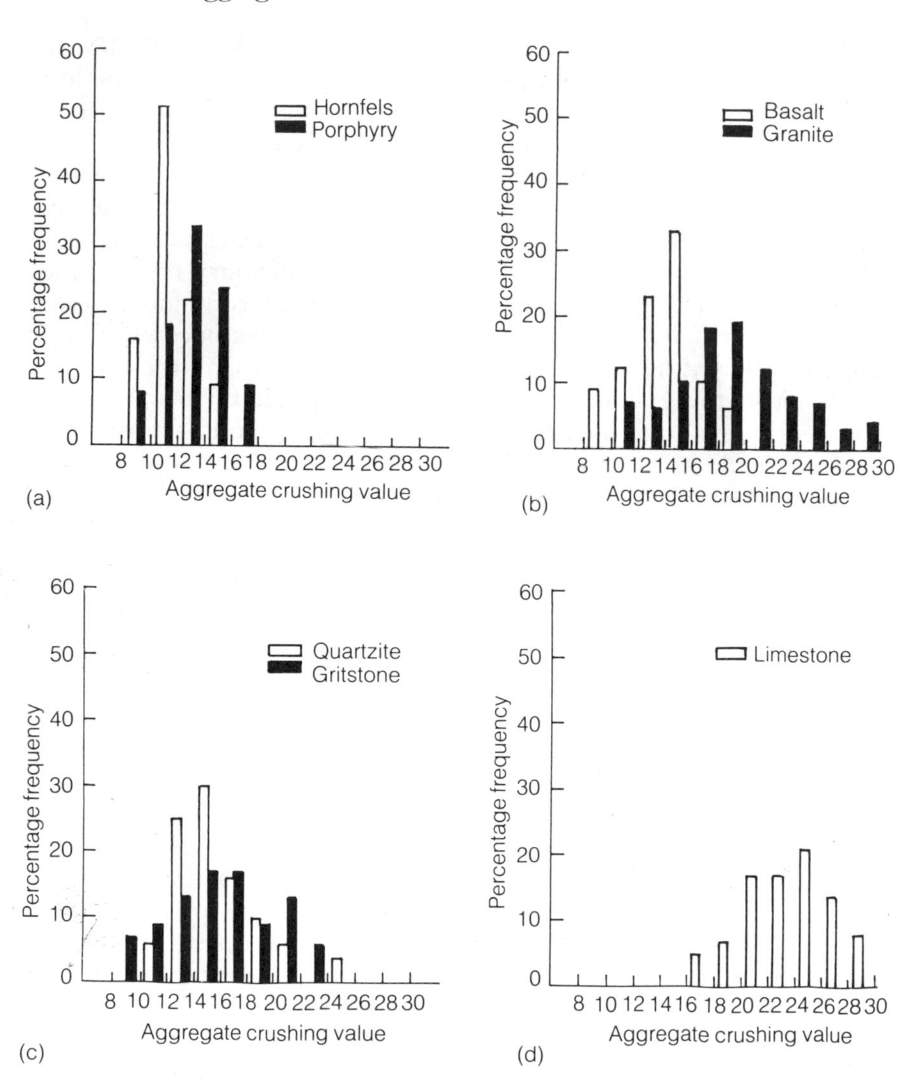

Figure 3.2 Aggregate crushing value of British roadstones in their trade groups: (a) metamorphic and igneous; (b) igneous; (c) sedimentary; (d) sedimentary. Data from *Road Research Road Note* 24 (Road Research Laboratory, 1959). Vertical scale = frequency of observations in above data.

more resistant to abrasion than gritstones (where the cement is more commonly calcite); and that basalts, despite a very different mineralogy, rank equal in this respect with granites. It seems probable that, in this area, grain-size and cement are more significant than the mineralogy of the grains.

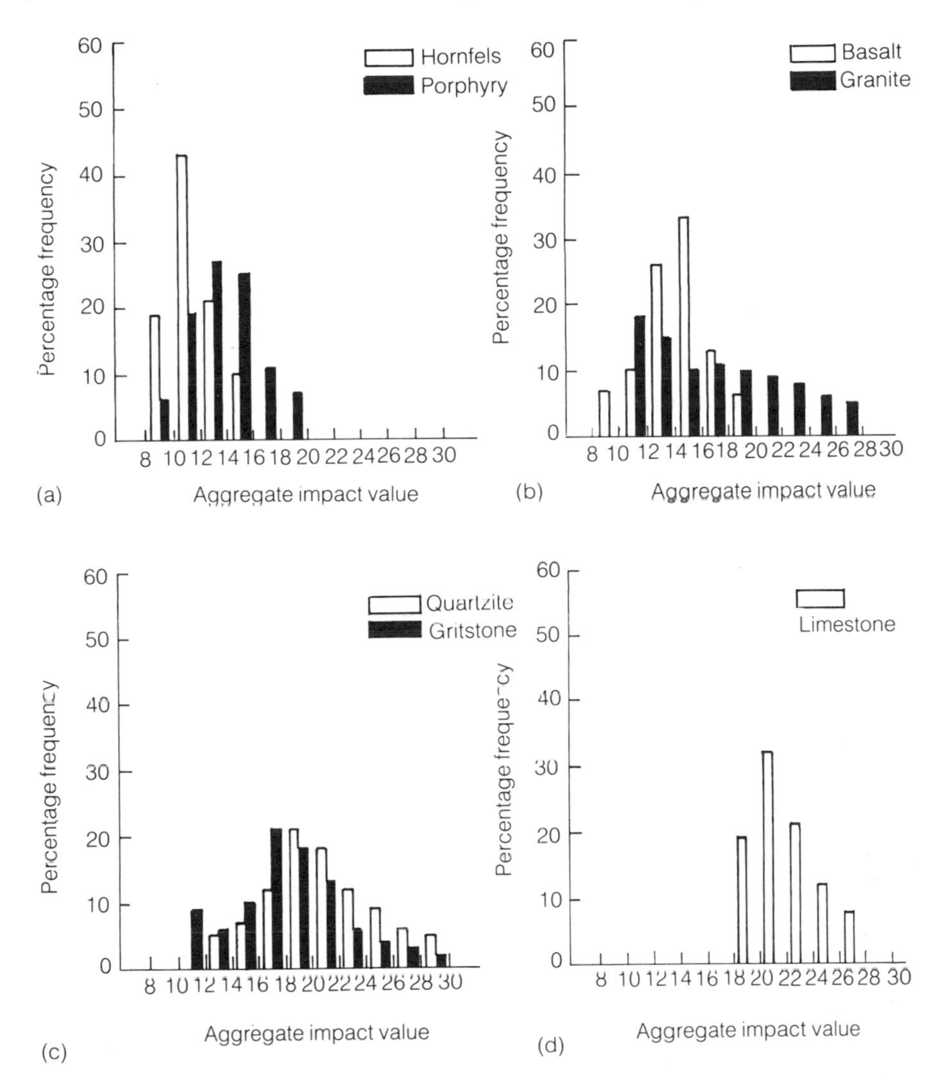

Figure 3.3 Aggregate impact value of British roadstones in their trade groups: (a) metamorphic and igneous; (b) igneous; (c) sedimentary; (d) sedimentary. Data from *Road Research Road Note* 24 (Road Research Laboratory, 1959). Vertical scale = frequency of observations in above data.

The **polished stone value** (PSV) is a measure whose importance lies in the recognition that, after a period of trafficking, some aggregates suffer a decline in their ability to inhibit skidding; this is related to the ability of these aggregates to lose their initial, rough surface, and develop a polish. A test value is derived in a machine which

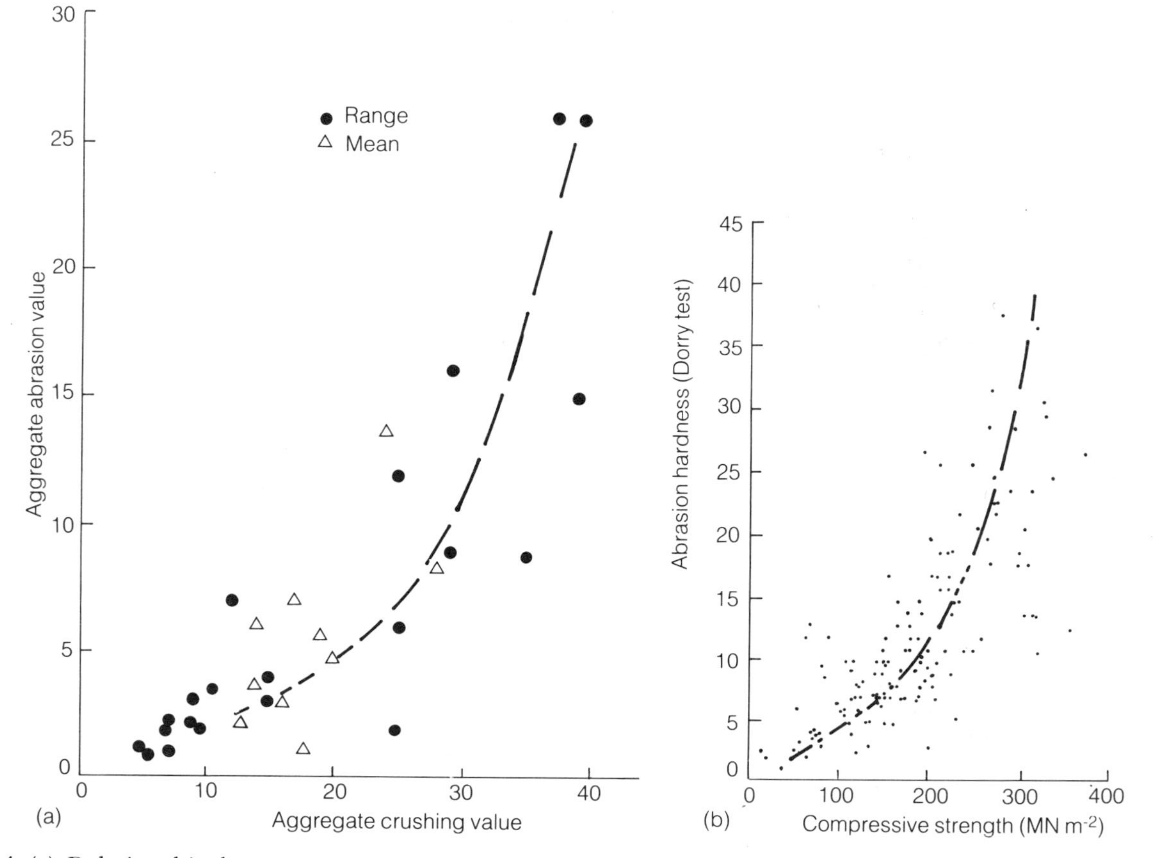

Figure 3.4 (a) Relationship between compressive strength and abrasion hardness after Kazi and Al-Mansour (1980). Data from Road Research Laboratory (1959); (b) Relationship between aggregate crushing value and aggregate abrasion value after Kazi and Al-Mansour (1980). Data from USBM, 1949–1956.

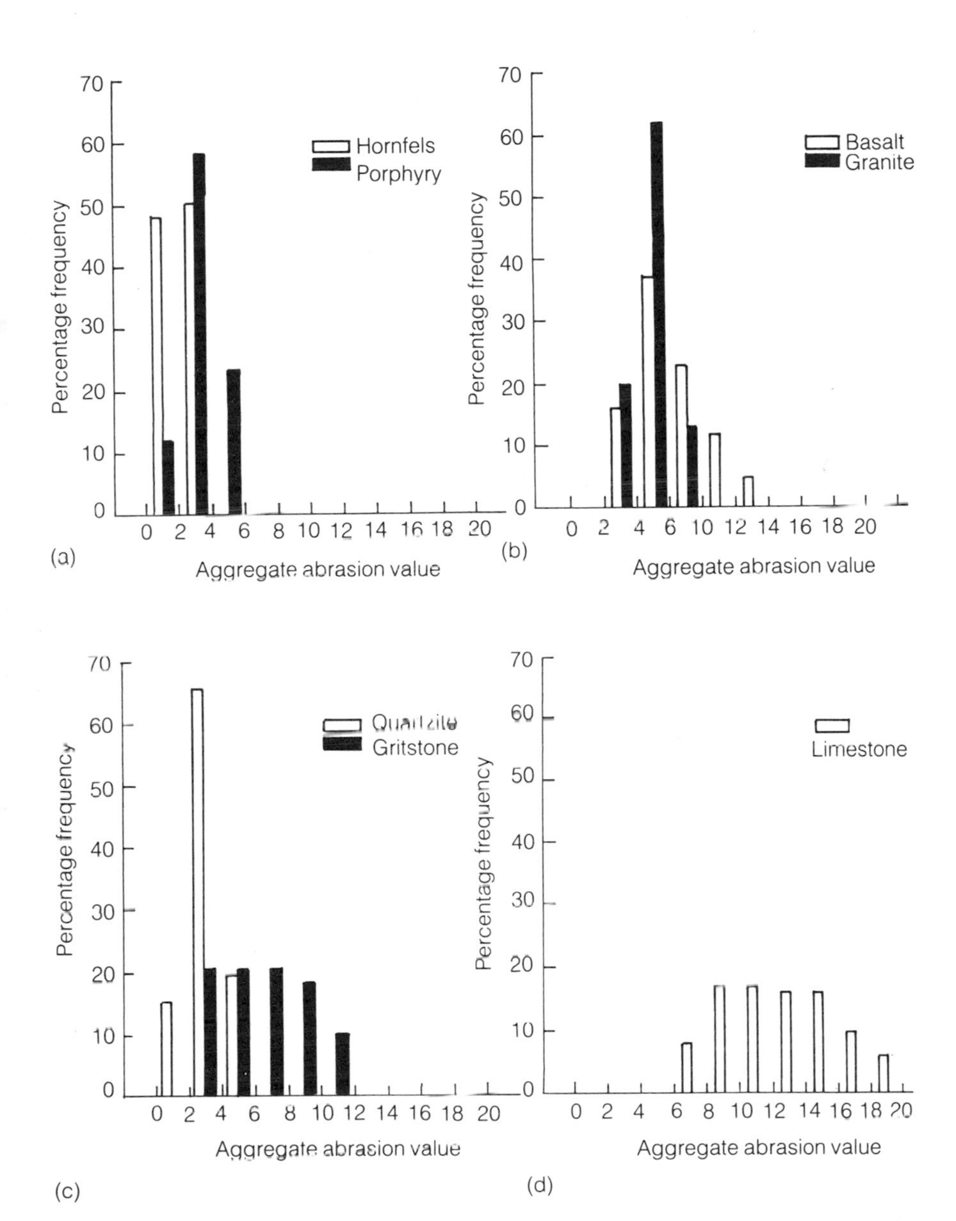

Figure 3.5 Aggregate abrasion value of British roadstones in their trade groups: (a) metamorphic and igneous; (b) igneous; (c) sedimentary; (d) sedimentary. Data from *Road Research Road Note* 24 (Road Research Laboratory, 1959). Vertical scale = frequency of observations in above data.

artificially accelerates this process, giving numbers in the range 20 to 72, in which the higher number denotes a better resistance to polishing. In British practice a PSV of 65 is regarded as the lower limit for a skid-resistant surface, but aggregates with these values are comparatively rare, and are usually required where the road presents a particular hazard. An extensive series of tests on British rocks has shown that the best surfacing materials from this point of view are certain greywackes from Palaeozoic sources; these are fine-grained rocks containing substantial quantities of each of the constituents quartz, chloritic clays and lithic fragments; they have suffered a moderate degree of metamorphism, and are thus dense and compacted, with an intergrown texture. They appear to owe their high PSV to the fact that each of the main constituents abrades at a different rate, so that a constantly rough surface is presented to vehicle tyres. This does not apply to all greywackes, however, and the relationships to mineralogy are complex; in general the best polished stone values are found in greywackes which have a high porosity, a high grain density, and more clay and lithic fragments, while higher quartz is disadvantageous. Quartzites and arkoses generally polish readily – they have PSVs of 50 to 55. Among non-sandstone rocks (Figure 3.6), the limestones polish most readily (PSV range 20–50), although the occurrence of sand in the limestone can increase this to over 65. Most igneous rocks – granites, basalts, gabbros and dolerites – lie in the range 35–65, but one hypersthene-dolerite from the Welsh borders exceeds 65 PSV.

There is an interrelationship between all these measures of mechanical properties; but because in natural rocks there are so many possible variables, the relationships are not simple. Within any one group of rocks, for example in the arenaceous rocks depicted in Figure 3.7, the values for impact and abrasion show a positive correlation; but these both have an inverse relationship to polished stone value. Thus many rocks which have a good polish resistance must be rejected for road surfacing because they break down too rapidly under abrasion or impact.

Flakiness is a measure of the shape of the aggregate particles. Many rocks show a linear or platy structure due to the parallel orientation of elongate or platy minerals; while mica and clay minerals are the most obvious in this respect, any mineral which shows elongation – for example, quartz clasts in a sandstone – can produce such a structure. This lineation can be imparted in sedimentary rocks during deposition (clast- or bedding lineation) or during diagenesis and metamorphism (cleavage or foliation). In igneous rocks it can be formed by flow structures during

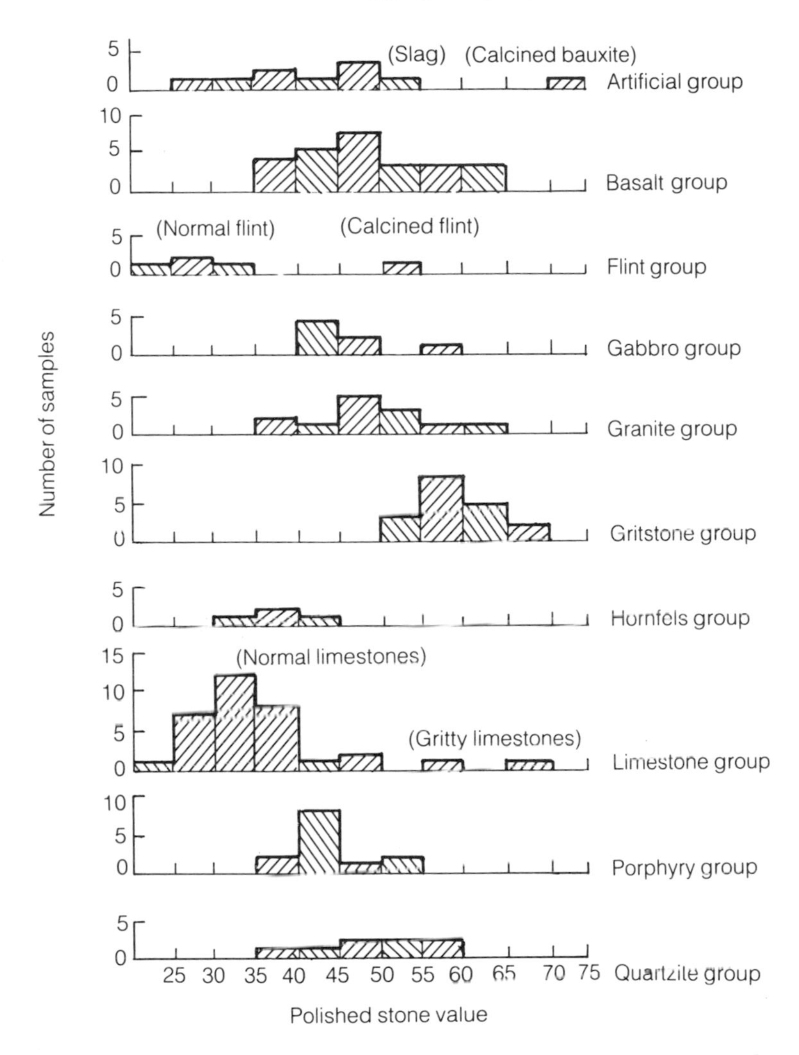

Figure 3.6 Distribution of polished stone values in different groups of rock (after Hartley (1970)).

emplacement or by sheet jointing during cooling. Subsequent closely spaced jointing can impart this to either igneous or sedimentary bodies. Thus it is a very common phenomenon, of which few rock-masses are totally devoid.

On crushing, these rocks break into fragments which are elongate and flattened rather than cuboidal (Figure 4.1). Quantification of this

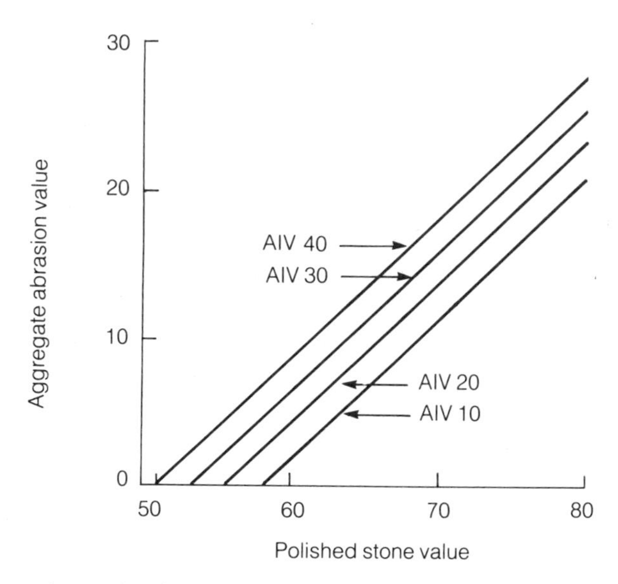

Figure 3.7 Relationship between PSV, AAV and AIV for British arenaceous roadstones (after Hawkes and Hosking (1972)).

phenomenon is by measuring the three dimensions of at least 200 pieces; the flakiness index (I_F) is defined as the weight percentage of fragments whose least dimension is less than 0.6 times the mean dimension: while the elongation index (I_E) is the weight percentage of fragments whose long-dimension is greater than 1.8 times the mean dimension.

The laborious nature of these measurements is justified by the importance of flakiness for the mechanical properties of the aggregate. It has been shown above (in this section and in Figure 3.1) that the strength of a platy rock, such as slate, is much less when measured normal to the lineation. An aggregate made from a platy rock will reflect the strength of that rock in its weakest orientation, especially as in any use – as on a road surface or as part of a concrete mix, the flaky particles will tend to align themselves. Flakiness thus substantially reduces the test-values of the aggregate. In Figure 3.8 is shown the worsening of aggregate impact value with increasing flakiness index from 0 to 100% – in granites from 17 to 25, in limestones from 15 to 24.

Resistance to weathering is difficult to measure, but is of equal importance to the other parameters. Aggregate used as a road surface is exposed to atmospheric weathering, which may be much accelerated by the breaking down of the aggregate by traffic loads. Weathering

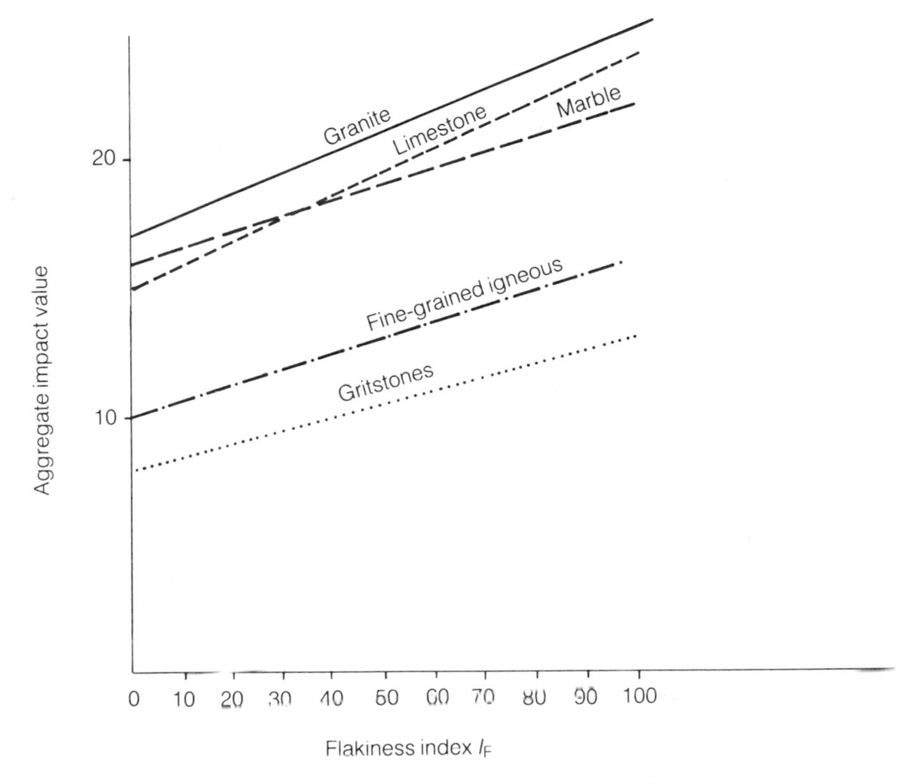

Figure 3.8 Aggregate impact values of different rock-types in relation to their flakiness index (after Ramsay *et al.*, 1974).

effects may also be manifested where aggregate is used as road sub-base, particularly where the surface is not totally sealed, and there is thus percolating water present; and where aggregate is exposed, or partially exposed, at the surface of concrete.

Of first importance is the necessity to ensure that the aggregate in these situations is not one in which the weathering process has already begun. The worsening of the mechanical properties on weathering is well recorded; but many quarry operators are unaware of the depth from surface to which this can extend, or the variability in this depth; and thus unsound weathered rock can be delivered to site together with the sound material. Visual observation by the quarry geologist, using criteria of open joints, clay-filled fissures, etc. is often adequate to prevent this occurring; but if quantitative confirmation is required the method of a micropetrographic index (as described above) can be used.

The degree with which even sound rock is susceptible to atmospheric weathering is variable, and attempts have been made to measure this. The slake durability test, as described in section 2.8, can be used to measure the effect of exposure to saturation; the Washington degradation test used in America is based on a similar procedure. In general they are only applicable to mudrocks, which would only in exceptional cases be used as aggregate. Stronger rocks, however, can be shattered by frost, or by extreme changes of temperature. Several measures, such as the ASTM C88 soundness test, widely used in America, attempt to simulate the effect of ice- or salt-crystals by alternate saturation of the aggregate with sodium or magnesium sulphate, and drying; the loss of weight of the aggregate after a standard number of such cycles is used as the measure of soundness. Alternatively, actual freezing and thawing may be used as this measure, as in the British TRRL frost susceptibility test.

Finally, a property of great significance in the use of coarse aggregate is its **particle size distribution (grading)**. Coarse aggregate is usually sold in a series of nominal size grades (e.g. 14 mm, 20 mm, etc.). These size grades are produced by screening processes, and the ability to produce these grades depends upon the size distribution present in the raw material (in natural aggregates – section 3.3), or the efficacy of the crushing process (for crushed rock – section 3.3). Screening can never be wholly effective, so that each nominal size grade includes some undersize and some oversize particles. Limits are set in most countries to the permitted over- or undersize; thus in Britain aggregate sold as '14 mm' cannot contain more than 15% retained on the 10 mm sieve or more than 7% passing a 6.3 mm sieve. There are further limits set for aggregate to be used in concrete production. It is, however, hard to find any scientifically based guidelines for the size requirements for particular applications, and the selection of the grade of aggregate used appears to depend largely on experience and tradition. The only point on which there is general agreement is the necessity to exclude 'fines' – variously defined, but mostly as 'passing 75 μm'; this size will include most of the clay minerals, which do have a deleterious effect on the setting of concrete, and on the coating of bitumen; but there are some rock formations which contain no clay, so that even the 'fines' need not be harmful. The construction industry thus relies on tests carried out on the final product – concrete, bitumen mixes etc. – rather than specifying the ingredients in terms of grading; the responsibility of the supplier is to provide a consistent product within defined limits of variation.

3.3 COARSE AGGREGATE FROM CRUSHED ROCK

Before the present century, most of the demand for coarse aggregate was met from the naturally occurring gravels; since many major cities were situated in large river valleys, the alluvial plains of the rivers were a natural and immediate source. Such sources of supply, however, were soon seen to be insufficient, and there has been a gradual replacement of them by crushed rock. In road construction crushed rock has almost entirely replaced natural aggregate, although the latter is often preferred for concrete. Nevertheless, in Britain today, quantities produced from natural aggregate and crushed stone are almost equal at around 100 million tonnes each per year.

As described above, an aggregate needs to meet a series of closely defined specifications; and there are a limited number of geological formations which can supply that need. Economic pressures have tended to favour large production units, producing many millions of tonnes annually, which need of course a correspondingly large base of raw material. Quarrying and processing hard rock is, despite the best efforts of the industry to contain it, a noisy and dusty process, and generates heavy traffic, and there are thus severe restrictions on environmental grounds – especially as 'hard rock' areas are also commonly regions of scenic interest and attraction.

In Britain the pattern is clearly defined (Figure 3.9), with the geologically young South-East being almost devoid of hard rock, and the north of Scotland generally too far from heavy population to make development worth while. The youngest rocks to provide aggregate are the Kentish ragstones – silicified limestones of Cretaceous age – which are not now greatly exploited, but interesting as being the nearest to the London conurbation. Apart from these, the most southerly rocks to provide aggregate, outcrop in the belt of Jurassic rocks extending diagonally across England from the Dorset coast to Yorkshire. These Jurassic limestones occur as a series of very large lenses (up to 200 km in length) at a variety of stratigraphical horizons from Middle to Upper Jurassic. Consistent limstone beds of good quality rarely exceed 30 m in thickness, and there are frequent zones of soft limestone and clay horizons. Locally, there are areas of higher quality limestone, but in general they have a low strength, and rather poor abrasion and polished stone values. Nevertheless, their location near the major centres of population has ensured their extensive exploitation.

Limestones of better quality occur in the Carboniferous; hard, well cemented, and often silicified, and occurring in consistent thickness

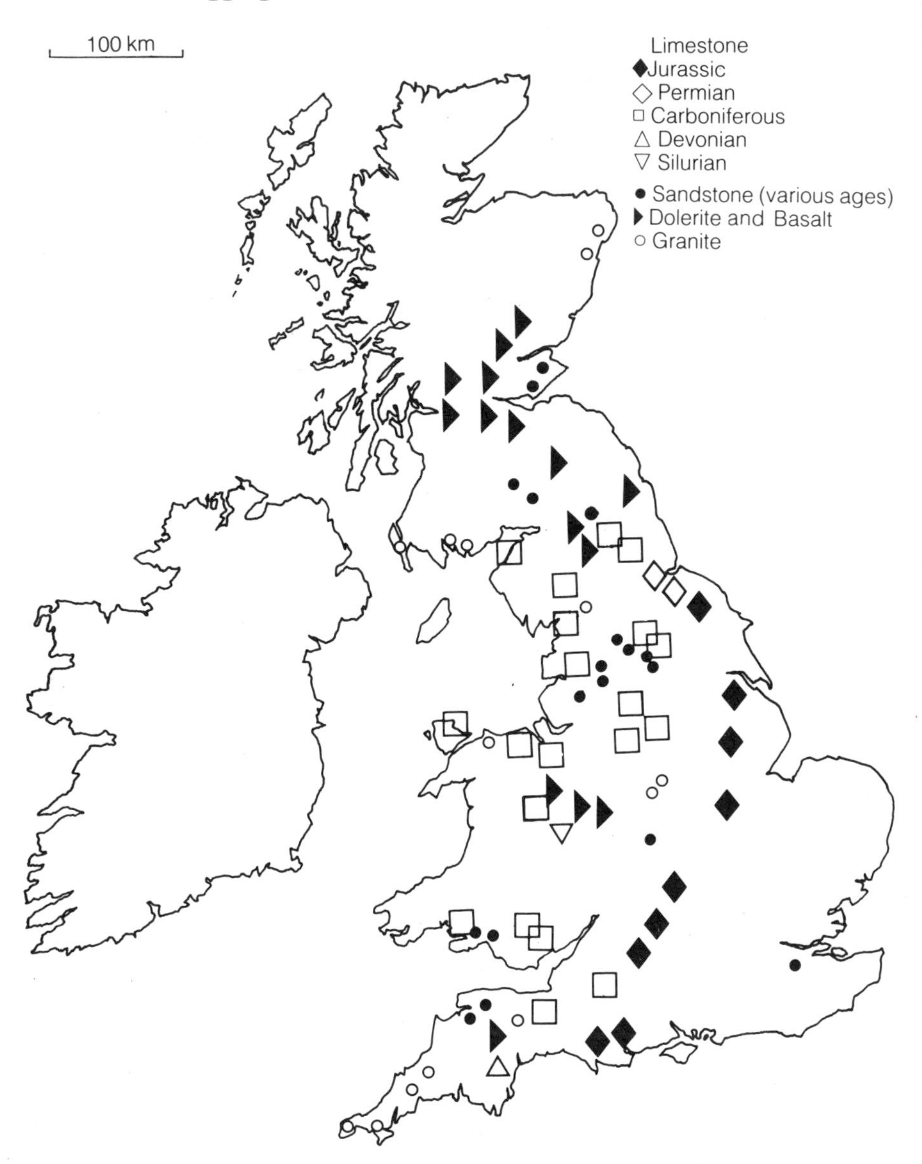

Figure 3.9 Principal crushed-rock aggregate producing areas in Britain.

often of several hundred metres, they are among the most extensively exploited minerals of the British Isles. Many very large quarries are found in this formation, whose outcrop frequently rims the major coalfield area: in the Mendip area of Somerset, in north and south Wales, and in the central and southern Pennines. In the latter area these limestones are frequently very pure, and exploitation for their chemical purity is in competition with that for aggregates. In the northern Pennines, and northwards to the borders of Scotland, the massive Carboniferous limestone facies gives place to a shaly facies with only limited limestone horizons; one of these, the Great Limestone, is locally over 100 m thick, and is exploited at many quarries. When the Carboniferous reappears in the Midland Valley of Scotland, the limestones have become too thin and shaly to be quarried for aggregate.

Some older limestones are quarried to some degree – for example the Silurian Wenlock and Aymestrey limestones of the Welsh borderland, and the Devonian limestones of south Devon. Altogether limestones form over 60% of the total crushed rock produced for aggregate in the British Isles.

Also in the north of England is the outcrop of Permian magnesian limestone, which extends along the east side of the Pennine hills from the Northumberland coast to the latitude of Nottingham. Considerably stronger and harder than pure limestone, this dolomite is the basis for a chain of quarries along its length. Again, there is competition for its use, since the magnesium content is valued for agricultural, refractory and chemical purposes.

The Carboniferous sandstones of the central and northern Pennines (section 2.4) have a disadvantage as an aggregate, since they contain much mica, which tends to be orientated along fairly closely spaced bedding planes or shaly partings. In consequence, the coarse aggregate tends to be flaky; and so it is not surprising that the output of these quarries for aggregate is almost matched by that for paving flags.

The Carboniferous rocks of northern England, and of the Midland Valley of Scotland, are intruded by a series of basic dykes and sills. Of the latter, the great Whin Sill of Northumberland is best known – this, with many others further north, and some basic lavas of related age, supplies a very large part of the crushed rock aggregate of northern England and Scotland. Dolerites and basalts are notorious for the ease with which they weather, their ferromagnesian minerals producing smectitic clays; for a long time the depth to which this weathering can extend from the surface was not recognized, and so weathered material was incorporated with unweathered, and this caused many failures.

Granitic rocks are quarried for aggregate in five separate areas of Britain; these are

1. the large granite batholith of south-west England, extending from Land's End to Dartmoor
2. a single granite pluton in Leicestershire; and geographically contiguous, but geologically unrelated dioritic rocks
3. the Shap granite on the margins of the English Lake District
4. a series of granitic bosses along the south-west coast of Scotland; these have an economic advantage in being situated along the coast, so that aggregate quarried here is directly accessible for sea-transport to the south-east of England
5. the Aberdeen and Peterhead granites of north-east Scotland

All these granites were intruded after the latest tectonism in the area – they have thus suffered little deformation, and foliar structures, which would give rise to flaky aggregates, are uncommon.

In respect of the supply of crushed rock aggregate, Britain is a microcosm of north-west Europe (Figure 3.10). In the north, the very extensive Fenno–Scandian shield area has an abundance of gneisses and other highly welded metamorphic rocks; and in this zone, as well as in the adjacent Caledonian mountain belt of Norway, there are large post-tectonic granites. Because of its distance from the heavily populated zones of central Europe, this large resource remains comparatively unexploited; the quarries tend to concentrate on producing dimension and ornamental stone, whose higher value can better bear the cost of transport. In north-central Europe generally the situation is as in Britain, with the population living on Mesozoic and younger rocks, and supplied with aggregate from the Palaeozoic massifs which locally penetrate the Mesozoic cover.

Thus the Ardennes massif supplies Belgium and northern France with aggregate from Devonian limestones and sandstones. The Carboniferous limestone facies similar to that of England extends along the northern borders of the Ardennes massif, and its use for aggregate is in competition with demand for it as agricultural lime. It is not found in the Eifel and Rhenish massifs, which thus supply the industrial Ruhr from Devonian sources. Supply is augmented in the Rhineland region, however, by Pleistocene volcanic ashes and lavas. Some of the ashes are suitable for use with minimal crushing, but there are also basaltic lavas which are fully indurated. The higher Rhine valley is also supplied by quarries in porphyritic lavas and some sedimentary rocks in the old massifs on either side of the Rhine rift, and from the Pleistocene Kaiserstuhl volcanics in the rift itself. Further east, the Harz massif contains a substantial granite body, now largely in East Germany.

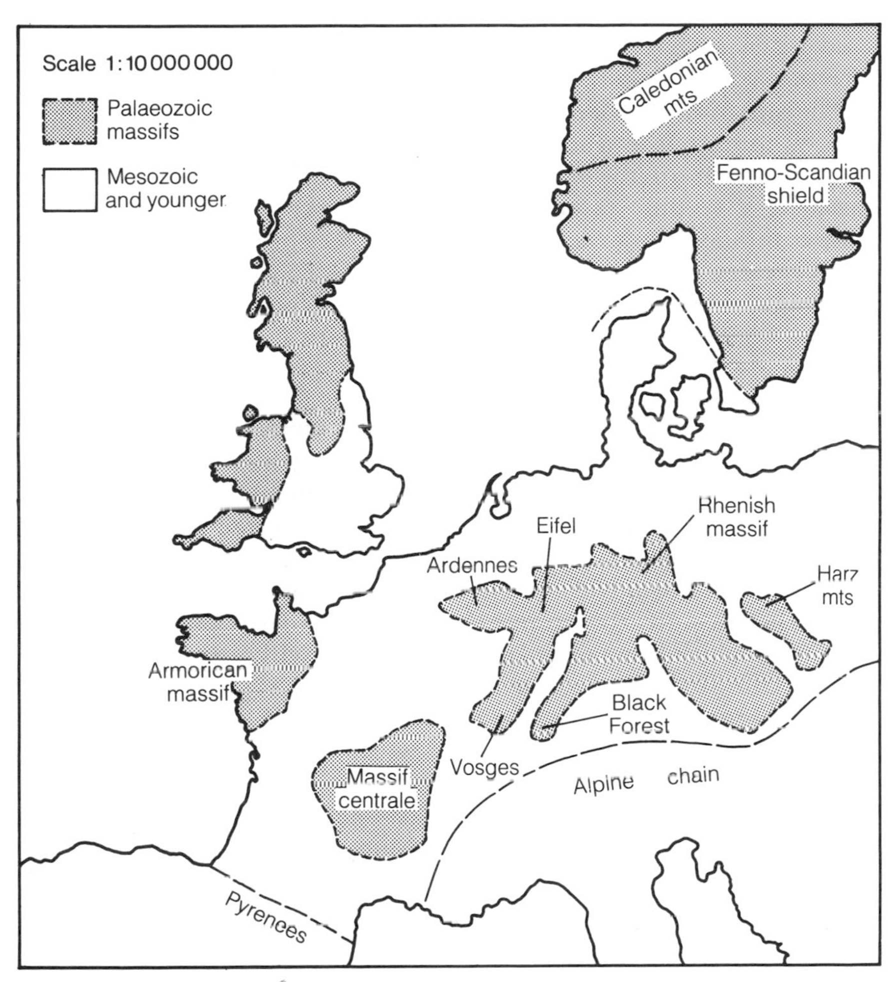

Figure 3.10 Map of north-west Europe showing distribution of (a) Palaeozoic massifs and (b) mainly Mesozoic and younger terrain.

Further south, as the Alpine front is approached, the largely Mesozoic terrain is dominated by limestones, and over most of southern and Mediterranean Europe this is effectively the only crushed rock aggregate available. Some of the older limestones, caught up in the Alpine orogeny, have been converted to marble, and where not suitable for dimension stone, are used extensively for road surfacing and for concrete – as for example, in many areas of Greece.

The situation in Mediterranean Europe, which is wholly dependent on limestone as a source of crushed rock aggregate, is replicated in many parts of the world. Large areas of the Middle East, for instance, have only rather weak limestones to use in concrete, and this, combined with the disintegrating effect of saline groundwater in a very dry climate, has led to some disastrous failures in modern buildings. This same aggregate situation is found over almost the whole of southern and western North America, and some extensive areas of Australia.

Another zone which presents problems is the lateritic weathering zone of the present-day hot, humid regions of the world. The enormous depth of weathering, which may reach hundreds of metres, means that over large tracts of country there is no hard rock available; or any that is available is so deeply rotted that it makes a very imperfect concrete. Many such regions are in the poorer and least developed countries; and it is no exaggeration to maintain that if such countries are to acquire the living standards enjoyed by the developed world, one of the major problems will be the provision of an adequate supply of coarse aggregate.

While aggregate sources may be widely variable, methods of extraction are similar throughout the world. Nearly all crushed-rock aggregate is won by open-pit extraction. The first stage in the process normally requires explosives; a line of shot-holes is drilled at a distance back from the quarry face, and explosive charges inserted and fired. The pattern of blast-holes, the amount of charge used, and the timing of the blast must be carefully regulated to ensure the maximum efficiency of the explosive, the production of appropriate-sized blocks, and the creation of a new, stable quarry face. It is thus a highly skilled operation. The experienced explosives operative will use the natural discontinuities in the rock – joints, bedding etc. – to the best advantage; these are usually evident in the existing rock-face, but the geologist has a responsibility to warn when there are likely to be changes behind the face which are not apparent – as, for instance, when a major fault or flexure is being approached.

The **primary blasting** operation will produce blocks of various sizes, some of which may be too large to be sent directly to the crusher. If this is so, the larger blocks may be reduced by **secondary blasting** – small charges being inserted into shot-holes drilled in the blocks, and fired on the quarry floor. Alternatively they may be broken up by a steel ball raised and dropped by a crane.

The next stage is **crushing**, generally carried out at a large fixed plant, although for small-scale operations mobile plant may be sited within the quarry. There are many types of crusher available, but they may be summarized into four main categories (Figure 3.11). The **jaw crusher** crushes the rock between two steel plates, one of which is fixed while the other is moved inwards against it by an eccentric on the driving wheel; thus as the material is crushed to smaller sizes it moves further down the jaws, finally escaping at the base – the maximum size can then be controlled by varying the

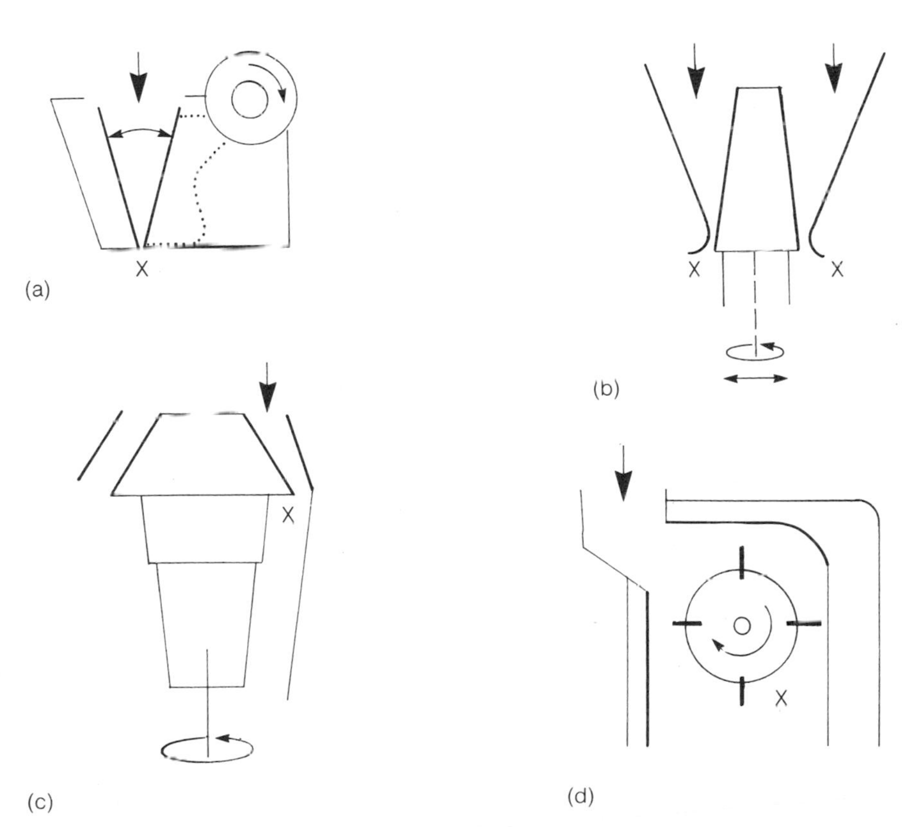

Figure 3.11 Diagrams to illustrate the principles of the four main types of crusher: (a) jaw crusher; (b) gyratory crusher; (c) cone crusher; (d) impact crusher. Crushing surfaces in solid black. X, exit aperture.

aperture at the base of the two plates (Figure 3.11, X). The **gyratory crusher** (Figure 3.11(b)) consists of a fixed inverted cone into which the rock is fed from the top; within this is a another narrower cone which rotates about an eccentric axis; thus the space between the two crushing surfaces is regularly opening and narrowing. Somewhat the same principle is used by the **cone crusher** (Figure 3.11(c)), but here the two conical crushing surfaces are sub-parallel. The **impact crusher** (Figure 3.11(d)) works on an entirely different principle, with rock entering a chamber in which is a rapidly rotating disc, primary, secondary, and frequently tertiary crushing is normal. hammers, against the wall of the chamber, and against one another, and thus reduced in size. Reduction from quarry-run rock in blocks of perhaps a metre or more in diameter, down to the sizes needed for coarse aggregate, cannot usually be achieved in one stage, and so **primary**, **secondary**, and frequently **tertiary** crushing is normal. Generally the primary crusher is a jaw or gyratory type, but there are many combinations possible.

It will be realized that forces applied to a block of rock are different in each of the four types of crusher; in the jaw crusher the block is subjected to compression from two directions, while in the gyratory type there is an element of tangential stress as well. Crushing by impact is an entirely different mechanism. It is not surprising, therefore, that the shape and size of fragment produced by each of these processes is different; this is controlled in the first instance by the structure of the rock itself, but can be modified to some extent by selection of the crushing process, and by the control of such variables as the rate of feed, the spacing of the crushing surfaces and the setting of the exit aperture. Thus a rock which may tend to produce a flaky aggregate in a jaw crusher, may be reduced to more cuboidal shapes in an impact crusher.

Although the maximum size of reduced fragment can be controlled by setting the exit aperture of the crusher, there are inevitably many sizes of fragment below this dimension which are produced. The pattern of size distribution shows considerable variation. Figure 3.12 shows the size distribution of the product of a jaw crusher with different feeds and different aperture settings. Of course, the maximum size is controlled by the setting, but the similarity of the three curves, despite the very different nature of the rocks – granite, slag and basalt – suggests that the crushing process may be more important than the mechanical properties of the rock. This is apparently assumed to be true by plant manufacturers, who invariably supply data on their product without reference to the rock type being crushed. Certainly, the different types of crusher behave differently,

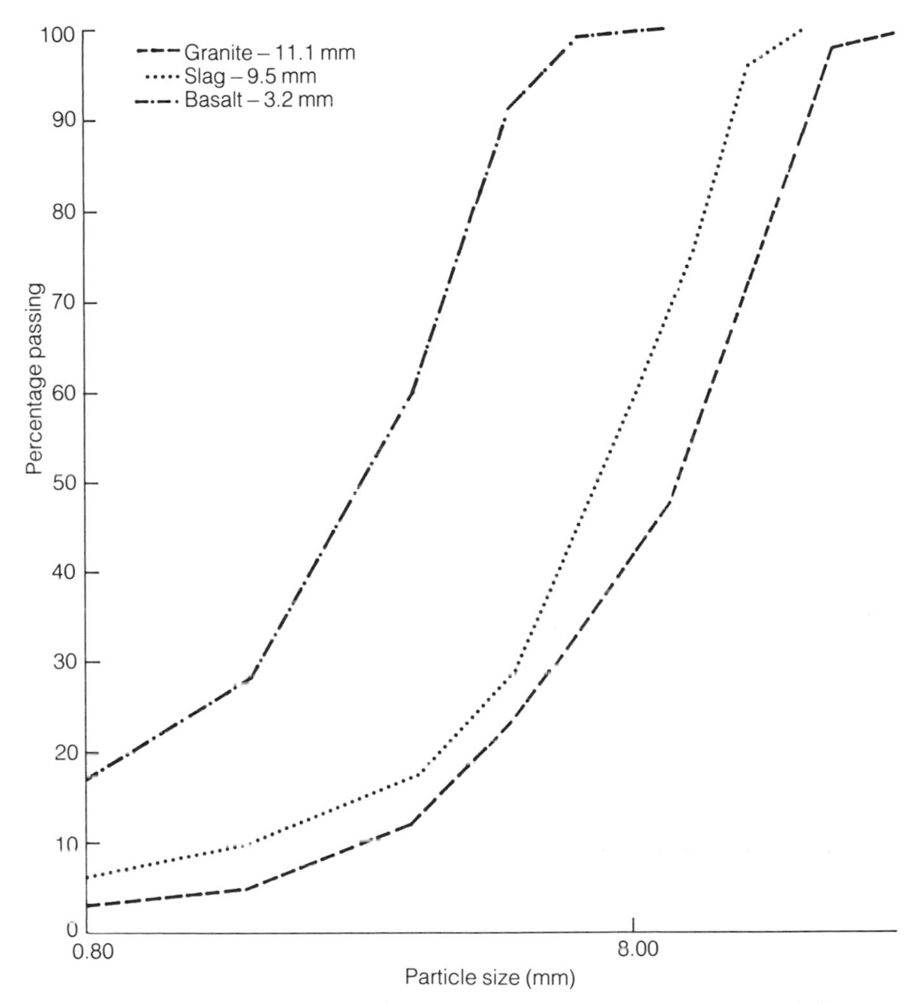

Figure 3.12 Size distribution of product of jaw crushing at three different settings, and for three different feed materials. Data from Brown (1963).

as is shown by Figures 3.13 and 3.14. In finer products (Figure 3.13), gyratory and cone crushers produce a similar product – since their action is somewhat similar this is to be expected; but the product of jaw crushing is markedly different. This difference is even more apparent at coarser grades (Figure 3.14).

One feature originating in the rock itself is, however, of great importance – that of **abrasiveness**. The amount of wear on the crushing surfaces of any type of crusher is very great, and the cost of replacing these is high; thus an abrasive rock such as, say, a silicified

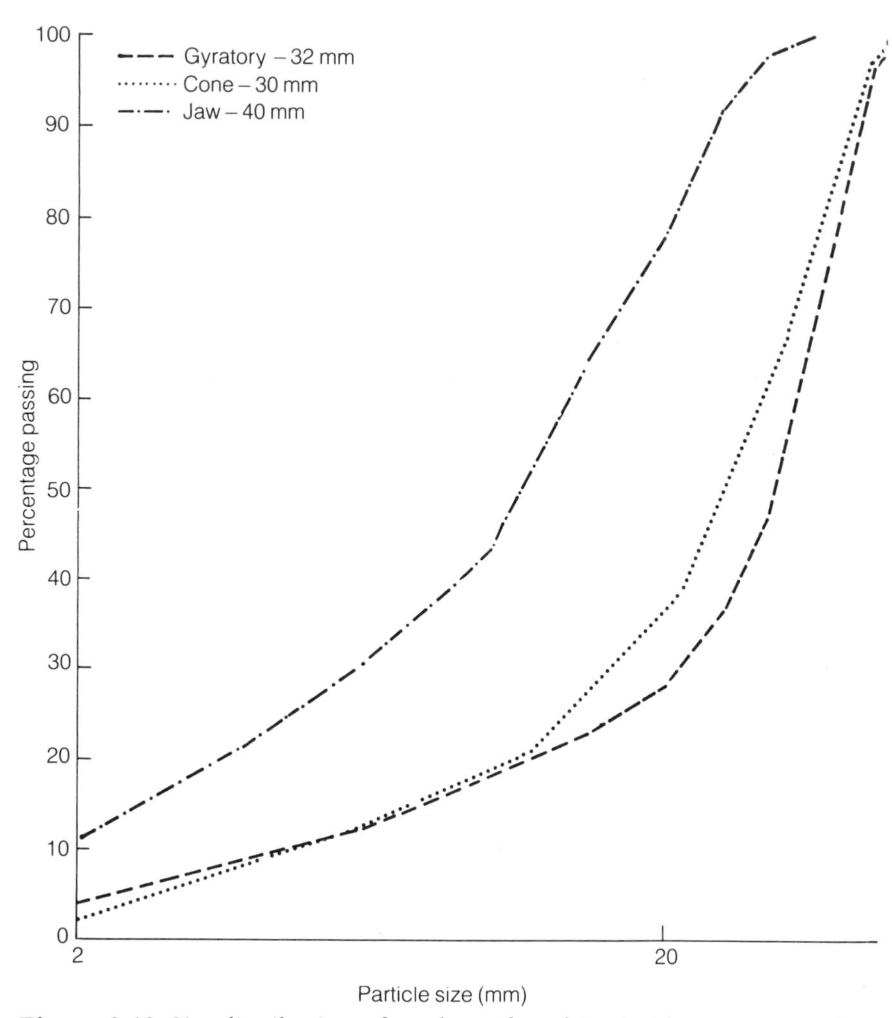

Figure 3.13 Size distribution of product of crushing in (a) gyratory crusher (32-mm setting); (b) cone crusher (30-mm setting); and (c) jaw-crusher (40-mm setting). Data from manufacturers.

limestone, is much more expensive to produce as aggregate than a softer, unsilicified material. Abrasiveness in a rock – the ability to abrade – is not always the same as abrasion resistance; so crushing an abrasive rock may not necessarily result in the production of a better quality aggregate.

The final process in the production of crushed rock aggregate is **screening**, which separates the aggregate into oversize or undersize fractions; such screens may be wire mesh, parallel bars or perforated

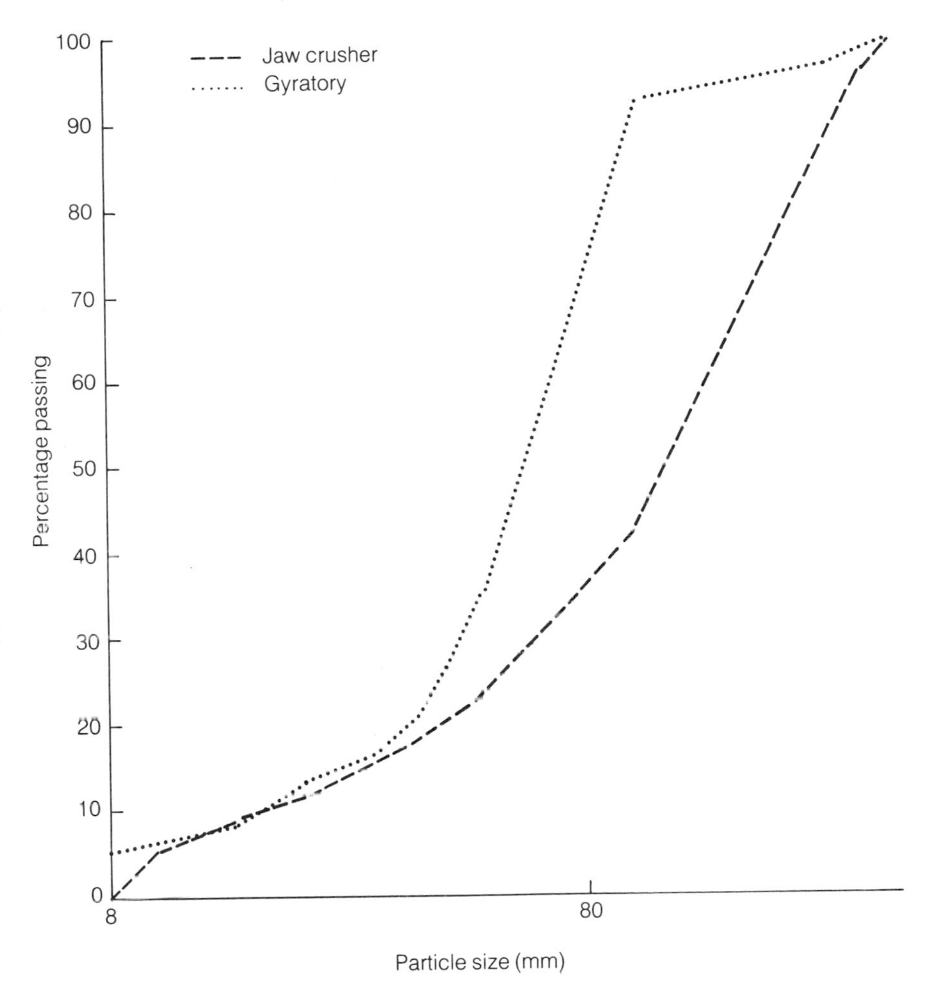

Figure 3.14 Size distribution of product of crushing in (a) jaw crushers; and (b) gyratory crushers at similar settings. Data from manufacturers.

plates, and made of steel, rubber or polymer. It is not practicable to use dry screening methods for material below 3 mm in diameter; but it is the main process for size separation in the range of sizes of coarse aggregate.

3.4 NATURALLY OCCURRING AGGREGATES

Sands and gravels is the usual name for a variety of sedimentary deposits, which contain rock fragments (clasts) of a size that can be used directly, or with minimal processing, as coarse and fine aggregate. They occur in a variety of geological situations; sometimes in sizes (**boulders** or **cobbles**) which require some reduction before they can be used; sometimes as **pebbles** which can be used directly to provide coarse aggregate. Sometimes natural sorting processes have separated the size grades; sometimes all sizes, cobbles, pebbles, sand, silt and clay occur together.

The method of formation of such deposits determines their characteristics. The **sedimentary source area** and its **weathering regime** determine the original petrology and shape of the clasts; for example, if the clasts originated in a limestone terrain subjected to an arctic climate, the clasts would be angular and consist entirely of limestone; if in a metamorphic terrain with a tropical climate, as a varied suite of clasts, of which only the most chemically resistant would survive. During **sedimentary transport** abrasion and attrition round the clasts, eliminate the softer ones, and sort them – the degree to which these processes occur being dependent partly on the mode of transport – river, glacier, or sea – and partly on the distance they are transported. Finally, **depositional processes** effect further sorting, and determine the shape and variability of the final deposit. Thus a full understanding of these deposits, and their geological history, is essential if the geologist is

1. to locate new deposits, and extension of known ones
2. to properly interpret borehole records, so that his evaluation of quantity and quality is correct, and
3. to devise the most efficient extraction pattern.

Many sand and gravel deposits were formed during Pleistocene and Recent times, and are thus a reflection of the region's latest geological and climatic history. The influence of the Pleistocene glaciation has been profound, and has been a major factor in ensuring that the temperate zones of the world today have had an adequate supply of aggregate to create their concrete cities. There are other deposits formed earlier in geological history which are sufficiently unconsolidated to break down to sand and gravel with ease, and which were laid down under older and different regimes.

Since their mode of origin is so important a determinant of their properties, it is possible to use a genetic classification. They can thus be described as residual, colluvial, alluvial, marine or glacigenic.

Residual deposits are those which are formed *in situ*, that is, they have not been transported from their source. Mechanical and chemical weathering processes can break up the surface of a rock mass to a very substantial depth, and this weathering mantle can be dug for coarse aggregate. In many equatorial regions, where weathering processes are particularly intense, as in many parts of Africa and South America, this is the commonest source of aggregate. In gneissose and granitic terrains, where chemical weathering has been active, the more reactive minerals are reduced to clay, which is then washed out by rain and groundwater, leaving a loose mass of angular rubble of the harder quartzose materials. Graded, and sometimes washed, this can provide a good quality aggregate.

The only British example is the 'clay-with-flints' of south-eastern England. This was formed by the long-extended solution of the Cretaceous chalk, where percolating water has dissolved the calcareous material, leaving a residue of the insoluable parts – the flint concretions and the clay. Where the clay content is low, and particularly where there has been some limited reworking of the deposit, these flint gravels have provided a small source of aggregate for local use.

An important group of residual deposits, formed in regions where there is a high evaporation of water from the surface of the ground,

Figure 3.15 Duricrust overlying alluvial gravels, Oranjemund, Namibia (photograph: author).

and consequent upward migration of groundwater, are the duricrusts (Figure 3.15). These form in the first instance as pebbly, concretionary surfaces, which then unite to form a continuous crust which may reach several metres in thickness. It may then require some crushing to reduce it again to aggregate size. Such deposits form an important source of coarse aggregate in regions where often no other source exists – as in many parts of the Middle East. The composition of the duricrust is dependent upon the soluble content of the groundwater, which in turn depends upon the underlying geology. Calcrete (calcitic duricrust), and dolocrete (dolomitic duricrust) can both be used as aggregate; but the all-too-frequent association of gypsum, halite and other soluble salts makes many occurrences unsuitable for concrete.

The term **colluvium** is a useful label (roughly corresponding to the colloquial English **head**), for deposits which have emplaced by mass-movement – e.g. scree and talus slopes and the like. They are, of course, petrographically similar to the rock from which they were immediately derived, and the fragments are angular and widely variable in size. They tend to occur banked up against steep slopes, and their upper surface is defined by the angle of repose of the constituent materials, so that they are often of relatively small thickness. The cold periglacial phase which accompanied the advance of the Pelistocene glaciers produced many such deposits in southern Europe, which have since become vegetated and stabilized, and form a source of aggregate – as for example, in the southern Rhône valley in France.

The **alluvial** plains of major rivers are an important source of aggregate throughout the world; but much depends on their recent geological history. At the present time large rivers such as the Thames, the Seine, the Rhine and the Mississippi flow so slowly in their lower reaches that they are capable of transporting only fine sand and silt. However, during the Pleistocene glaciations, and particularly during the melting phases, their discharge and velocity was vastly enhanced, and they laid down substantial bodies of coarse gravel.

During the Pleistocene, the alternation of cold and warm periods produced rises and falls of sea-level; in times of low sea-level the rivers cut deeply into the bedrock, or into pre-existing alluvial deposits; at times of high sea-level the rivers deposited their gravel in these channels. Thus in a river such as the Thames, the gravel sequence can be complex (Figure 3.16). The lowest sea-level, at the time of the maximum advance of the ice, was well below that of the present day, so there is a deep, gravel-filled channel which is not

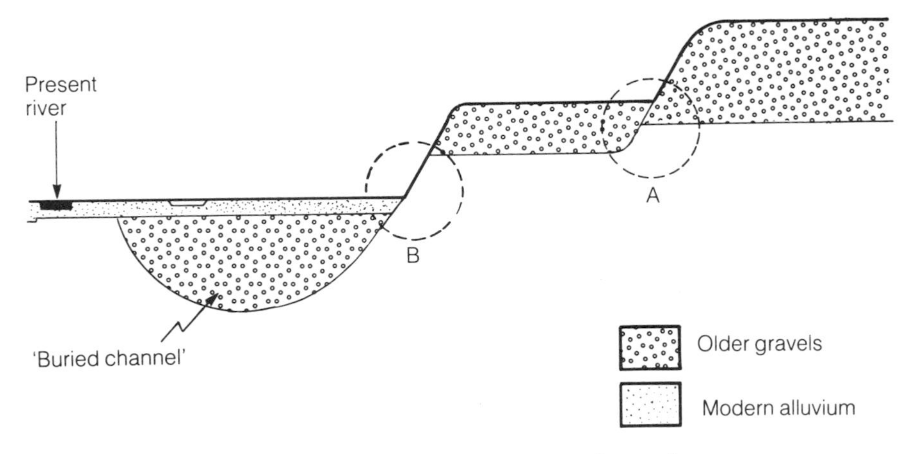

Figure 3.16 Diagrammatic section across river valley and terrace sequence.

always coincident with the course of the present river. Above this level is a complex of river terraces with various relationships with one another. It will be seen that, in Figure 3.16, the interpretation of a gravel exploitation in areas A and B might present difficulties in interpretation unless the full geological situation was understood; and since the different gravel terraces have different properties, this may be of substantial economic importance.

In Britain, the Thames Valley gravels have always been a major resource. A typical size distribution curve is shown in Figure 3.17(a), the total aggregate ('as dug') curve shows that there is a variation between the finest material in which nearly 60% is sand, and the coarsest in which some 22% is sand, and the remainder is gravel up to 50 mm in diameter. Separating the coarse fraction gives Figure 3.17(b), showing an even gradation which allows the production of a range of graded agregates. The gravels in the lower Thames valley are also much valued for their quality, since much of the stone consists of hard, rounded flint pebbles which have developed a tough outer skin or patina; these are very durable and non-reactive with cement. Originally derived from the chalk, it seems that these pebbles first formed part of the marine Tertiary sequence, and were then re-excavated before being incorporated in the Pleistocene gravels; other hard pebbles are derived from the glacial deposits immediately to the north. In the upper Thames valley, however, these hard pebbles are diluted by an influx of softer Jurassic limestone pebbles, since this rock forms much of the catchment area of the river in its higher reaches.

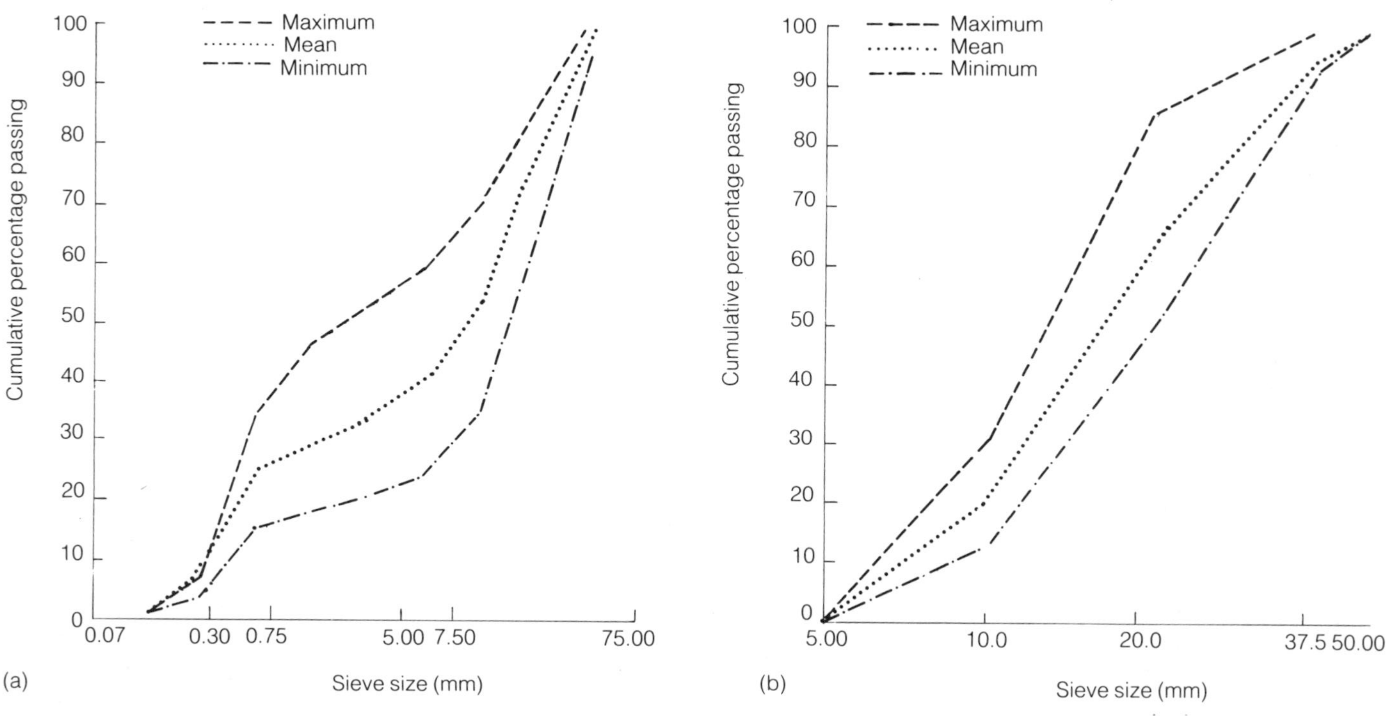

Figure 3.17 Size distribution of Thames Valley gravel: (a) total aggregate, coarse and fine, 'as won'; (b) coarse aggregate >5.0 mm.

A very different picture is presented by the deposits of a river in the north of England; here there is a buried channel (Figure 3.18(a)) filled by a pebbly sand of very uniform composition; this suggests a large river flowing into an estuary. This is overlain by a terrace deposit of markedly different composition (Figure 3.18(b)) – very coarse gravels suggesting a torrential river flow.

These two examples show how very different alluvial deposits may be from one another; there is also always very considerable variation within the alluvial deposit. River deposition is an intermittent phenomenon, connected with the deposition and re-erosion of the banks of the meandering stream. Within the deposit, therefore, gravel is deposited as a series of lenses, within and around which changes of grain-size distribution are rapid. It is thus rarely possible to discern a regular pattern of size distribution within the deposit.

In regions where large rivers drain nearby mountain ranges, and particularly where rainfall may be heavy, but markedly seasonal, the river may be braided, and the deposits spread over a wide river-bed which is largely dry for most of the year. This situation applies to many of the rivers of southern Europe. In these circumstances the river bed itself is quarried, often selectively for different grades of gravel, but sometimes, as for instance in the lower Rhône valley, very systematically. It is often assumed that this resource will replace itself at the winter floods, but whether total replenishment of some of the very large extractions of this kind actually occurs seems somewhat doubtful. It is characteristic of such deposits that there is a great range of size, with every size from boulders to sand being present; in such deposits the oversize is often crushed to produce the desired grades. Since deposits such as these often lie close to the source of the rock, and the clasts have thus not travelled far, the degree of rounding is often low, and softer rocks have not been eliminated by abrasion and attrition; the quality of the stone may not therefore be high.

Marine deposits have very different characteristics from those formed on land. In the cycle of sedimentary processes, all rock fragments derived from erosion of the land will eventually reach the sea. Viewed as an agent for the production of valuable sand and gravel, the marine environment has two important characteristics. First, bodies of sea-water generally move slowly, so that their ability to transport clasts larger than fine silt and clay is limited to certain precise locations. These locations are

1. the littoral zone, from a few metres below low water level to a few metres above high water mark, and
2. the floors of estuaries and shallow seas.

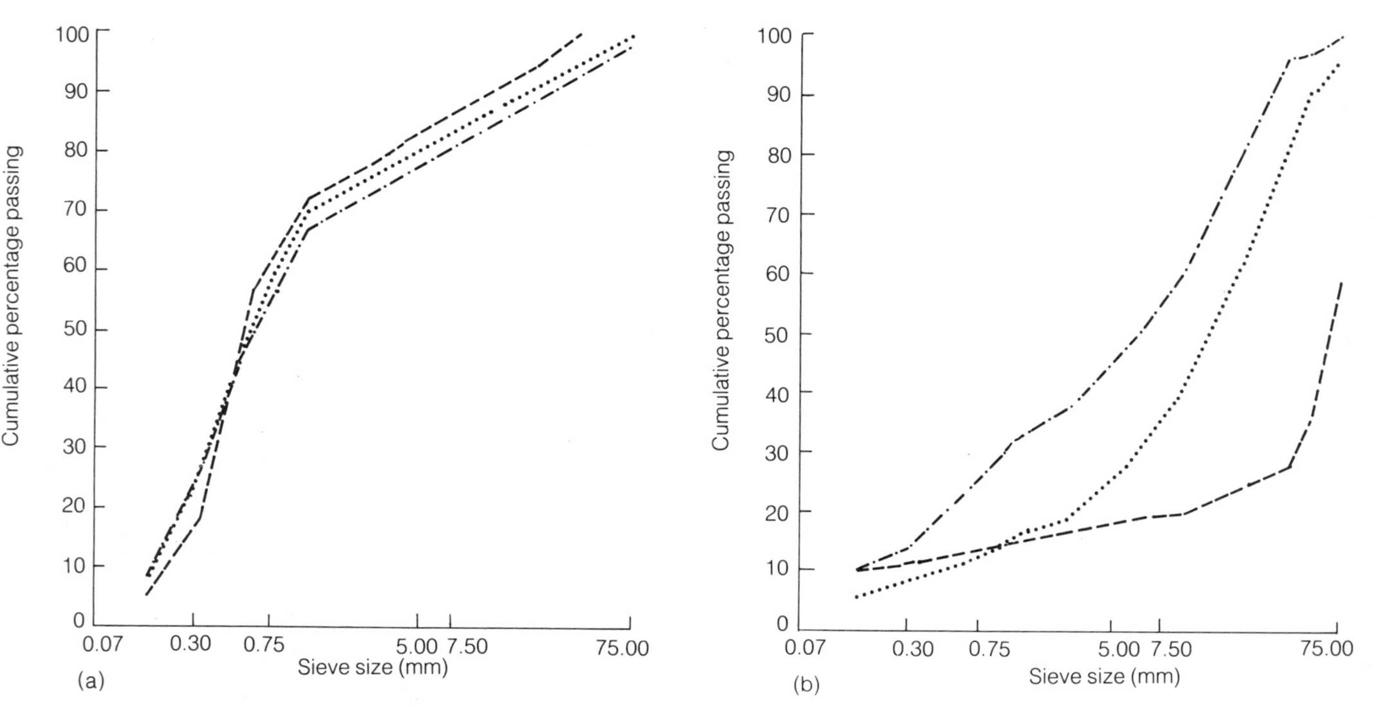

Figure 3.18 Size distribution of glacial and alluvial gravels, from a valley site in north-west England: (a) glacial channel fill; (b) alluvial terrace. Total aggregate, coarse and fine, 'as won'.

In the littoral zone, the main agency of transport is the energy of breaking waves, which is responsible for the formation of beach sands and shingle ridges along exposed shorelines. In shallow seas, the daily movement of the tides is confined, and the tidal current velocity sufficiently enhanced to allow transport of pebbles and sand.

The second characteristic of marine transport is that, where it does occur, it involves constant movement. Whereas a pebble moving down a stream moves only a few metres at a time, and then may remain unmoved for years, the pebbles on a shore, or on the sea-bed, are in constant motion. The process of pebble attrition is thus very rapid; softer rock-types are rapidly worn away, and the gravel soon comes to consist only of the very hardest material. Marine-won gravels are thus often of very high quality.

Another limiting factor in the occurrence of marine sands and gravels is the source of supply to the sea. Many of the largest rivers in the world – the Niger, the Amazon, the Mississippi, etc. – bring only clay and silt to their mouths, so no supply of coarse material is available in these regions. In this respect, the northern hemisphere temperate zone has been fortunate, because the recent glaciers dropped large volumes of morainic material, containing all size grades of clast, on to the shores and beds of the fringing continental seas. It is the reworking of this glacial material by marine processes that has created the shore ridges, beach sands and gravel banks of this region. It must be evident, at the same time, that this is a finite resource, and there is some evidence in the erosion of beaches and shore ridges, that the supply is dwindling.

Another factor in the exploitation of marine sands and gravels is their undoubted role in coastal protection. The presence of a coarse pebble beach at the base of a cliffed coast is efficient in reducing coastal erosion; and many low-lying areas are protected from inundation by a coastal barrier. In many coastal regions in the past the beaches and shingle ridges of the shore have been excavated for aggregate – a useful and easily available source of such material, and one which manifestly was constantly renewed. But awareness of the potential dangers of increased erosion has caused strict controls of this activity in most countries, and usually is now only permitted where the beach ridges are in a stable or 'fossilized' state – as, for example, in the Dungeness region of southern England (Figure 3.19).

The movement of sea-bed gravel and sand by tidal currents is perhaps less clearly understood, and it is impossible to prove or disprove whether the removal of a few million tonnes of gravel several kilometres offshore will affect the coastal defences of the

Figure 3.19 Aerial photograph of the Dungeness region, Kent, England. A succession of shingle-ridges built up by shoreface deposition during the late Pleistocene. Extensive 'wet' sand and gravel workings on west side (photograph Aerofilms Ltd).

neighbouring land, or destroy the habitat of important benthic fauna, or affect changes of shipping channels – but all such matters need properly to be considered before sea-bed exploitation is allowed.

Since different processes of sediment transport and deposition are active in the sea, it is to be expected that the size distribution of marine deposits will be different from that of alluvial areas. Figure 3.20 illustrates this: the coastal ridge gravel shown in Figure 3.20(a) displays a very wide range of variation, from pebbly sand of which 98% passes the 600 μm sieve, to coarse gravel in which only 10 % is less than 5.0 mm. The separated coarse aggregate from this (Figure 3.20(b)) also shows a wide variation – this can be a desirable attribute since it allows the operator to produce a wide range of gradings. The high percentage of very large material, however, demands that some crushing of the oversize is carried out if material is not to be wasted. The range of variation of sea-bed aggregate (Figure 3.20(c) and (d)) is less, and shows a somewhat different pattern.

An advantageous feature of many marine sands is the low silt content – especially in dredged aggregates, where much of the fine material is washed out by the dredging. Disadvantageous features are the presence of salt and shell. Salt cannot be incorporated into concrete without disasterous effects, and so sea-dredged aggregate needs to be brought ashore, drained and preferably washed by fresh-water before use. Shell fragments are of course the remains of the marine molluscs; in coastal aggregates they are less frequent, because the violent attrition which produces beach ridges tends to wear down all but the most robust; but some sea-bed aggregates contain shell in abundance. Being soft, and composed of calcium carbonate, they lessen the strength, and sometimes produce undesirable reactions within the concrete.

With the decline of on-shore resources of sand and gravel, offshore dredging has been regarded as a solution to the supply problems; and extensive dredging is in fact carried out from gravel banks in the North Sea and English Channel by neighbouring countries. While undoubtedly this is an important resource, the material produced is by no means ideal, and the long-term effects on the adjacent coasts, and on the fishing and marine life, have yet to be properly evaluated.

A further category of naturally occurring aggregates is that of **glacigenic sands and gravels**. The continental ice sheets which covered much of the north of Eurasia and north America during the Pleistocene transported prodigious quantities of boulders, cobbles, pebbles, sand, silt and clay from the north to the south. When the ice sheets melted, much of this load of debris was deposited *in situ* as boulder clay or till. This is an extremely heterogeneous deposit,

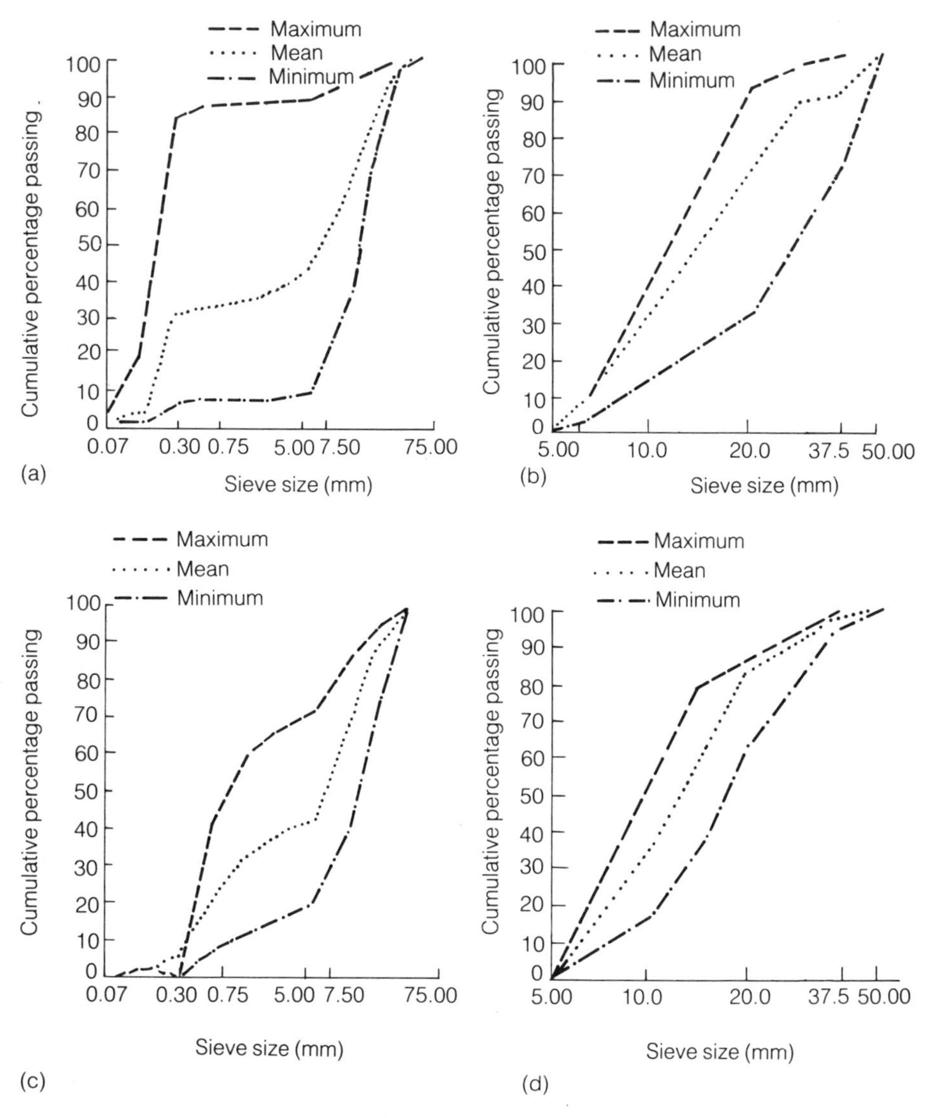

Figure 3.20 Size distribution of sands and gravels of marine origin: (a) total aggregate, coarse and fine, coastal ridge; (b) coarse aggregate, coastal ridge; (c) sea-dredged aggregate, coarse and fine, sea-bed; (d) sea-dredged aggregate, coarse, sea-bed. (a) and (c) show size distribution of material 'as won', (b) and (d) after processing to remove <5 mm.

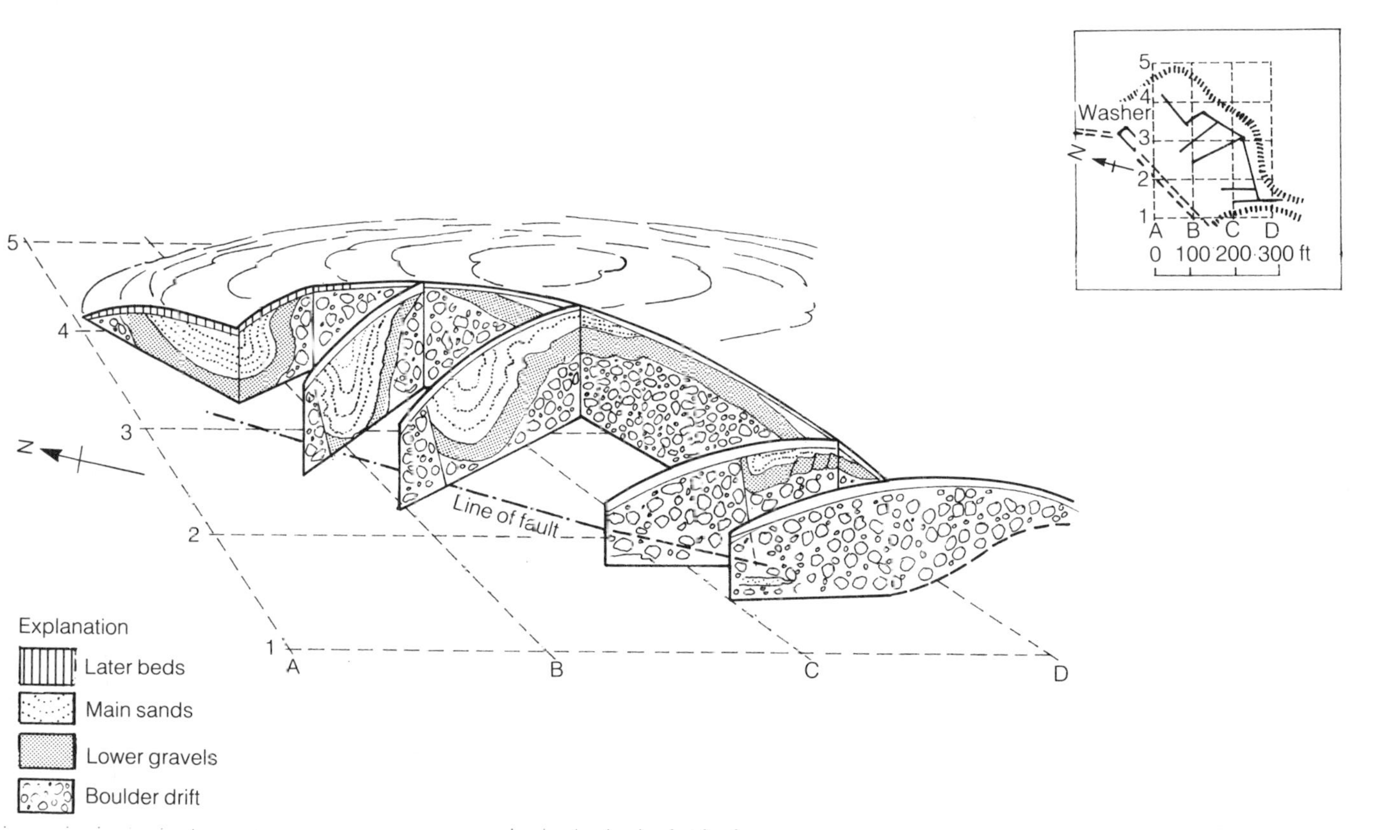

Figure 3.21 Diagram of a kame-moraine, Carstairs, Scotland (after Goodlet (1964), *Bulletin Geological Survey of Great Britain* No. 21.

usually containing such substantial quantities of clay that no processing will remove it; so, except where local conditions have produced a very sandy or clay-free till, it cannot be used to produce aggregate. However, during the retreat of the ice, large volumes of glacial meltwater flowed within the glacier and away from the ice front. This water was often very fast flowing, and was able to carry a large load of clasts out of the ice sheet; the sorting process of the flow carried the silt and clay long distances, leaving near the ice front extensive deposit of relatively clean sand and gravel – and these

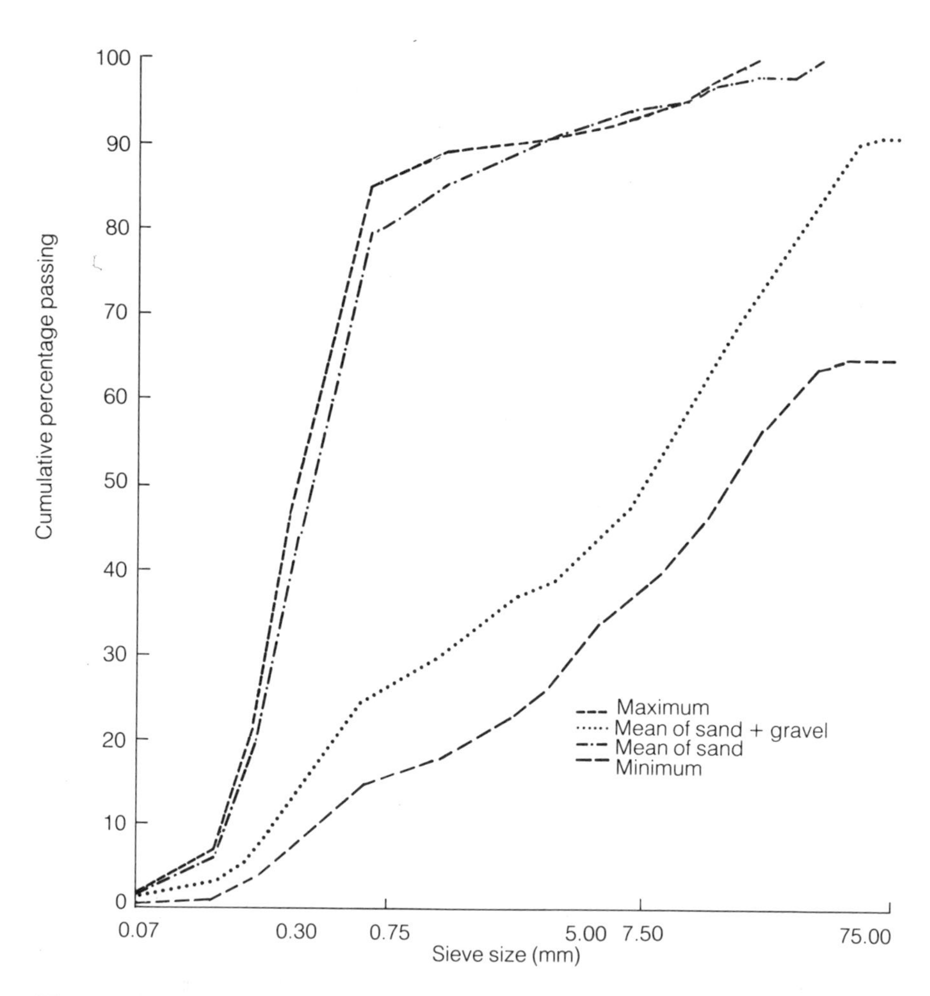

Figure 3.22 Size distribution of fluvioglacial sand and gravel deposit, Cheshire, England.

fluvioglacial gravels are important sources of aggregate in the recently glaciated areas.

Fluvioglacial gravels occur in several different configurations, often giving rise to conspicuous, and thus readily recognized, landforms. Thus kames or kame-moraines are hummocky mounds built up against the ice face (Figure 3.21); their complex structure is related to variable meltwater outflow and ice collapse. Eskers are sinous mounds built up at the mouths of ice-front streams; while delta fans are formed where the meltwater stream debouched into standing water. All these features have a complex internal structure, and a very careful appraisal and full understanding of the structure is needed before they can be properly evaluated.

Because the meltwater streams are often torrential, and deposition is therefore very rapid, fluvioglacial gravels are often poorly sorted, i.e. they show great variability in size range. Figure 3.22 shows such an analysis of one such deposit; the samples clearly fall into two groups – a sand and gravel, and a sand with only about 10 % pebbles – perhaps representing summer and winter flows of meltwater. Lateral and vertical changes of grading occur very frequently within such deposits, and local lenses of silt, and rafted masses of boulder clay, often make difficulties in quarrying.

There is also great heterogeneity in the rock types which make up the pebbles. In the process of ice movement, the glacier crosses over a wide variety of geological terrains, and picks up the loose surface material. The process of ice transport is not as effective as that of moving water in grinding down the softer rocks, so that fluvioglacial gravels often contain a substantial proportion of unsound aggregate. This is particularly true if the ice has moved over the outcrop of a softer formation in the last stages of its advance. Thus in Britain, fluvioglacial gravels south of the main coalfields often contain pebbles of coal, soft mudstone, and fireclay; in the English Midlands they have soft Jurassic sandstones and mudstones; in East Anglia pebbles of chalk. Thus the suitability of a fluvioglacial deposit for use as aggregate is often determined by the local ice movements, and a study of the known details of glacier formation and movement is often revealing in predicting the potential of such a deposit (section 4.4).

Fluvioglacial deposits are nevertheless a very important source of aggregate throughout the northern hemisphere. They are found over the whole zone of glaciation, extending in Europe northwards from its maximum extent at the latitude of the Thames and Rhine estuaries. They are particularly important in Scandinavia, where large alluvial tracts are unknown. Since the Alpine mountains also carried ice sheets at this time, fluvioglacial gravels are found, and exploited, in most of the river valleys radiating from the Alpine zone. Similarly, in north America the whole of the north-central region from Pennsylvania and Ohio northwards contains these deposits, while the local ice caps of the Rocky Mountains provided material which spread out over their eastern foothills.

While most of the unconsolidated sands and gravels are of Pleistocene or later origin, there are some more ancient deposits which are exploitable. Most rocks older than the Pleistocene have acquired some degree of cementation and compaction, so that the sands have been converted to sandstones, and the gravels to conglomerates. Sometimes, however, such rocks have remained sufficiently unconsolidated to be extracted without undue force, and broken down again by only minimal crushing to their original grain size. Such deposits, of course, reflect the conditions under which they were formed, and thus parallel in their characteristics the modern deposits laid down in the same way.

In the Anglo-Paris basin, a number of local pebble gravels accumulated as shingle ridges along the shores of the Tertiary seas; being mainly derived from the erosion of the widespread chalk, their pebble content is almost entirely the very hard flint. One good example is the Blackheath beds, whose extensive outcrop south and east of London was formerly much exploited; and similar examples are found in Belgium and northern France.

Older examples of 'fossil' aggregates occur in the Permian and Triassic rocks of Britain. During these periods, Britain was a mountainous terrain with a desert climate, and boulders and pebbles accumulated as detritus fans and alluvial cones at the margins of desert basins, or as deltaic fans in the transient lakes. Despite their age, they have undergone little cementation, and can be readily broken down to sand and gravel. One particular formation, now known as the Sherwood sandstone group, is extensively worked for this purpose in Nottinghamshire, Cheshire, Staffordshire and Devonshire. One feature of this formation, which adds much to its value, is that many of the pebbles are of a very hard quartzite; the origin of this durable rock is not known, although it is thought that it may have been derived from the erosion of the Armorican massif to the south.

Not only does the origin of a deposit determine its properties as an aggregate source, but also it has important repercussions on its extraction and processing.

Alluvial deposits often occur in areas where the water-table is high, so that in these cases the deposit is worked **wet**. Using a drag-line from the side, or a dredge placed on a floating pontoon, or sometimes a giant suction hose, and sand and gravel is brought up as a slurry. Most other kinds of deposits are **dry**, and are worked very much in the same way as brick-clays (Chapter 5). There are some advantages to wet working, in that some of the fine silt and clay is automatically washed out in the process of extraction; and the dredged material is often sufficiently liquid to enable it to be transferred to the processing plant by pipeline. In general, processing of natural aggregates is very simple, consisting mainly of screening to separate the size grades; if the original working was wet, it is usually screened in this state; otherwise dry screening is normal. If there are large amounts of large-sized cobbles in the deposit it is profitable to install a crusher to reduce these to saleable sizes.

3.5 CONCLUSIONS

The properties of a coarse aggregate, and therefore its suitability for road material, or for concrete, depend upon the petrography of the individual rock fragments of which is composed, and the size distribution, and shape, of those fragments. All these properties are in turn determined by the geological processes which formed the rock or deposit. By understanding these processes, and therefore their likely effects, the geologist can help to optimize production of consistent, good-quality aggregate. While total aggregate resources are quite inexhaustible, the need to find new aggregate sources within reasonable transport distances of their place of use is paramount; and the geologist needs to use all his or her skills to find these sources, and to ensure that they are exploited to their maximum efficiency.

4

Fine aggregate

4.1 DEFINITION AND USES OF FINE AGGREGATE

Fine aggregate is a cumbersome term for the material which in everday language would be called **sand**. Unfortunately, the word sand has implications which limit its usefulness. First, it implies a natural origin – on a beach, or in a dune for example – whereas much of the material used in industry is man made, by crushing rock, or even in some cases, crushing artificial products such as slag. Secondly, it carries the implication that the mineral grains are mainly quartz, whereas fine aggregates can be composed of different minerals, or of limestone, basalt, etc. Finally, whereas the grain-size classifications mostly used by sedimentologists (for instance, the Wentworth–Udden scale) place the upper size limit of sand at 2 mm, industrial usage places the limit of fine aggregate at 5 mm.

Fine aggregate has a variety of uses in the construction industry; the main ones are as follows:

1. as a major constituent of mass concrete, either formed in place, or as precast units
2. as a part of the 'unbound' section of roadway pavement.

In the above contexts it is always used in combination with coarse aggregate.

3. in the production of concrete products such as concrete **roof tiles** and drainage **pipes**
4. with cement, to produce **cement screeds** for floors, and internal and external **wall renderings**
5. with cement, and in the past lime, to produce **masonry mortars**
6. with plaster of paris to produce **gypsum plastering** for walls and ceilings
7. to provide the 'bound' section of roadway pavement, by mixing with bituminous compounds to form **tarmacadam** or **blacktop**

8. as a filler for trenches on which to lay electricity cables and some other main services
9. as a filter material for sewage and effluent disposal
10. as the primary constituent in the manufacture of **glass**.

In common usage are the terms **concreting sand** for usages 1 and 3, **building sand** for 4, 5 and 6, **asphalt sand** for 7, and **cable sand** for usage 8 – but these terms have no precise definition.

4.2 SPECIFICATIONS AND TESTING

With such a multiplicity of uses, it is not surprising that there is a wide variety of commercial specifications in existence. In almost all the usages indicated above, the role of the fine aggregate is that of a filler; that is, it fills in spaces which would otherwise have to be filled by binding material, or it provides stiffness to a mixture which would otherwise be soft and mobile. All the various binding materials in which fine aggregate is incorporated – cement, gypsum, bitumen – are themselves inherently weak, so that they act simply as a glue holding the aggregate grains together. Some binding materials are also prone to shrinkage during the dehydration following emplacement. The strength of the product depends therefore upon the strength of the aggregate, and the ideal situation is one in which the aggregate particles are closely packed, with a minimum amount of binder. There are also cost considerations; except possibly in the case of gypsum, the cost of the binder is much higher than that of the aggregate. In concrete products such as roof tiles and pipes, a cement:aggregate ratio of 1:3.5 is regarded as desirable, although not always achieved; but substantial cost-savings can be effected if this ratio is kept at a high level, and cement usage minimized.

Since the packing together of the aggregate grains is so important, it will be clear that their size distribution, and the shape of the particles, are important parameters. Size distribution is the easier to measure, and has received the greater consideration.

Grain size analysis is therefore a most important test procedure. Within the size ranges of fine aggregates, sieve techniques are those which are most appropriate for grain-size analysis, and these are widely used. In general the results are consistent and repeatable, but there are a number of factors which must be considered in the interpretation of results. These are as follows:

1. An individual sieve retains only those particles whose minimum cross-sectional area is greater than the size of the mesh opening. Thus elongate particles will slip through

the mesh even though their total size is much greater. In comparing two sands, therefore, it is important to make at least a qualitative judgement of the proportion of elongate particles.

2. Disaggregation of a sample is often not complete. The normal method of disaggregation is to apply gentle pressure, either by rubbing through the fingers, or with a rubber pestle. Too much pressure must be avoided, as it may break individual grains; and in an industrial context it is important that the test procedure should examine the sand as it will be when incorporated in the mix – thus total disaggregation may introduce fallacies. At the same time, incomplete disaggregation may also give false results, as aggregate particles of clay or silt may be retained on a coarser sieve, when they will be broken down to smaller sizes in the actual working mix. For these reasons, wet-sieving is often preferred to dry-sieving in sand used for cement mixes, but might be entirely inappropriate for an asphalt sand.
3. The sieving process can rarely be carried to completion. Even after very prolonged shaking, there is always a little more material passing the coarser sieves, although this clearly becomes less and less as shaking proceeds. The length of time that it takes to achieve acceptable accuracy cannot be precisely defined, as it is dependent upon the sand itself, and on the method and intensity of the shaking process, but for the most part, if standard procedures are adopted, meaningful comparisons between sands can be made on this basis. Care is needed, however, if international comparisons are to be made, since the standard laboratory practice varies from country to country.

In addition to the above, there are the usual hazards of laboratory practice – unclean and worn sieves being the commonest source of error. So sieve analyses always need to be interpreted with caution. A useful technique is to examine each sieve fraction under the binocular microscope; this always provides useful information on the presence of elongate grains, clay aggregates and the overall mineralogy of the sand, and is a useful check on the completeness or otherwise of the sieving process.

A measure derived from the grain size analysis is the **sand equivalent value**. In the grain-size distribution, one of the most important characteristics is the quantity of finest grain size, which is assumed to be clay. In sieve analysis it is defined as that passing the finest sieve, which may be 63 or 75 μm in mesh diameter. A quick method, defined in an American standard (ASTM 2419), employs a sedimentation technique, by which a liquid suspension of the aggregate, in a flocculating medium, is poured into a measuring cylinder and stirred. After the material has settled, it is usually possible to see a clear distinction between the sand below and clay above, and the percentage is estimated volumetrically.

Bulk density is an important characteristic of a fine aggregate. Since the closeness by which the grains are packed is the most important feature to be measured, a direct observation of this character is clearly important, and this can be provided by the measurement of the bulk density of dry compacted sand. A sand which had no voids would have a bulk density equal to the specific gravity of the mineral – which, if quartz, would be 2635 kg m^{-3}. Dry natural sands are reported as having bulk densities between 1400 and 1800 kg m^{-3}, representing void percentages of between 32 and 47%. Although an approximate correlation can be made between the bulk density of the sand, and the strength of concrete made from it, there are sufficient discrepancies in this correlation to show that the other parameter, particle shape, is at least as important as a determinant of concrete strength.

Particle shape is extremely difficult to measure in a sand grain, and no wholly satisfactory method has yet been devised. Although measurements under a microscope are of course perfectly possible, they are extremely time consuming, and most methods used depend upon visual comparison. It is recognized that there are two distinct elements in grain shape – **sphericity** and **roundness**, as follows.

(a) Sphericity

Sphericity is defined as the nearness of a grain to a sphere; it can be described in terms of the relationship between the three diameters of a grain – the longest (x), the shortest (z) and an intermediate (y) – and a classification based on the relations of these three has been established (Figure 4.1). This recognizes four shape classes, based on two ratios – a flatness or flakiness ratio (p) = z/y, and an elongation ratio (q) = y/x. The equant class (B) has its complete expression in the cube or the sphere (where $x = y = z$); grains depart in shape from this by being flatter or more elongate. Flatness without elongation produces flaky grains (Class A: $x = y > z$). Elongation without flatness gives prolate or rod-shaped grains (Class D: $x > y = z$). A combination of elongation and flatness gives rise to bladed grains (Class C: $x \neq y \neq z$). While it is possible to set mathematical limits

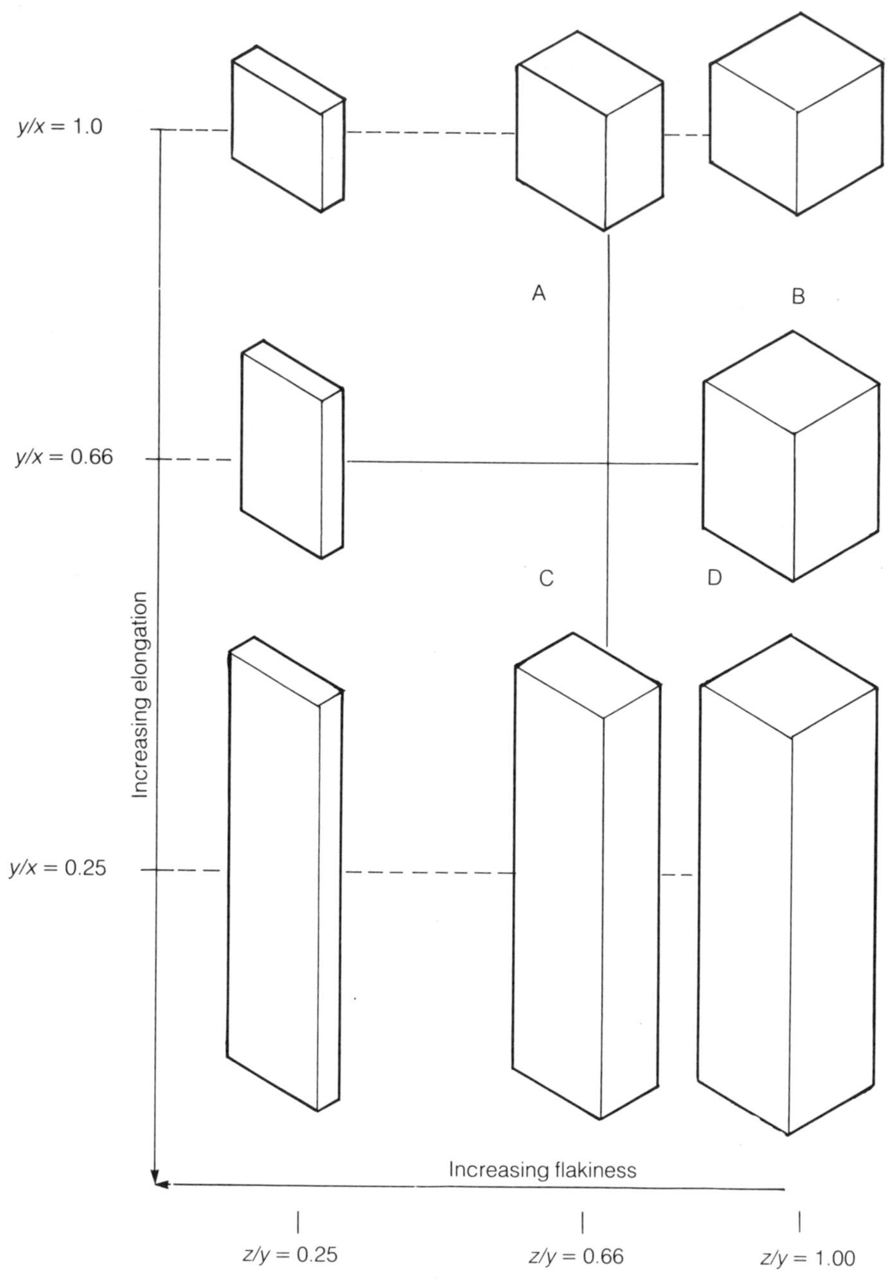

Figure 4.1 Description of grain shape (after Zingg (1935)). Grains are placed in four categories (A, B, C, D) on basis of proportions corresponding to the axial ratios y/x (elongation) and z/y (flatness or flakiness); x, longest; y, intermediate; z, shortest.

to each of these classes, most studies do not go beyond placing grains in one or other category by qualitative, visual appraisal.

(b) Roundness

Roundness is a measure of the degree of angularity of a sand grain. Again, visual comparison with a printed chart (Figure 4.2) is the only practical method; and on this basis sand grains can be placed in six categories, from 'very angular' to 'well rounded'.

A full study of the effects of roundness and sphericity on the technical properties of a sand has yet to be carried out. It is known that particle shape influences the packing of grains, and therefore must modify the influence of particle size in producing strength in concrete. The well-rounded grains of some natural sands are known to produce greater workability in mortar sands; at the same time they reduce the adhesion between cement and aggregate, and thus can impair performance in high-strength concrete.

As with coarse aggregates, there is a consensus that results of sieve analysis are presented in the form of a cumulative curve of percentages, using a logarithmic primary scale; although there is no general international agreement on whether it should be based on 'percentage passing' or 'percentage retained'. There are difficulties with sieve sizes, however. The imperial-measure-based British Standard sieve numbers are in desuetude, and replaced by diameter measures in microns or millimetres; increasingly American data are

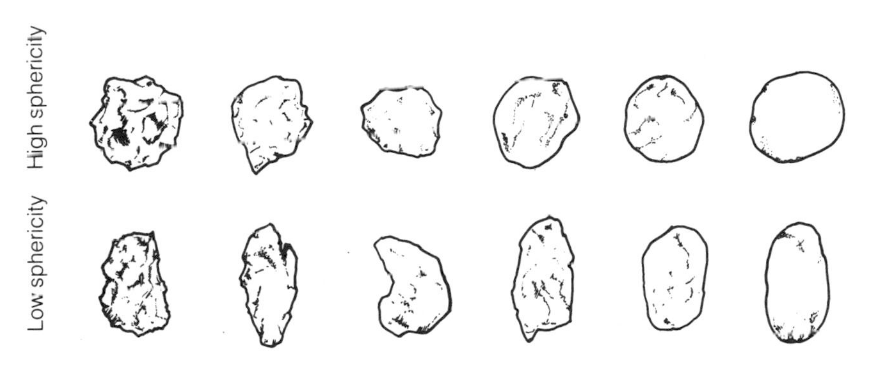

Figure 4.2 Categories of roundness for sand grains. A grain of high and low sphericity is shown for each category of roundness (after Pettijohn *et al.* (1973) Springer-Verlag, Berlin).

reported similarly. An intended International Standard has been totally ignored, since it uses a completely different range of sieve sizes. It is unfortunate that a large body of information, which could be of value in the industrial field, and contained in the very extensive literature of sedimentology, is invariably reported in φ-units (a measure defined as $\log_2$ diameter), or in derived statistical measure (φ-mean, φ-sorting, etc.) and thus difficult to transfer to the industrial system. In addition, the specifications for different end-uses demand different sieves – in Britain, for instance, the sieve sizes used for asphalt sand are different from those for concrete or mortar. In carrying out a sieve analysis for a sand whose end-use has not yet been determined, it is well to ensure that the range of sieves used covers all possibilities.

4.3 MINERALOGY OF FINE AGGREGATE

In medium to high latitudes of the globe, and especially in regions which have recently been glaciated, fine quartzose sands are usually abundant. In the process of transport by glaciers, by glacial melt-waters, and then by rivers, the softer minerals, such as feldspar, are abraded away; and the existence of the still waters of lakes, or of slow-flowing rivers, allows the accumulation of bodies of fine-grained sediments in which the major mineral is quartz. By contrast, in low latitudes, where chemical weathering is dominant, the laterization process removes the silica from the source rocks, so that most of the recently deposited sediments are argillaceous, and fine aggregate is difficult to find. Different again are the areas of recent mountain building, as for example most of southern Europe and western North America; here the potential source rocks are highly tectonized rocks of clastic origin, in which argillaceous rocks predominate, or limestones and dolomites, so that the weathering and transport processes produce little fine quartz; moreover, the high velocity of flow of most of the rivers carries fine sediment out of the area, often into the sea.

In addition to sources of fine sand of recent and Quaternary origin, some more ancient sediments are sometimes sufficiently poorly consolidated to be dug and used directly as aggregate – for example many Tertiary and Cretaceous sands in north-west Europe. Such sands are usually of marine origin, and include authigenic or other minerals in addition to quartz.

While the individual grains of fine aggregate are usually monomineralic, there are some instances of grains of this size being sufficiently complex to take part in the same reactions as the coarse aggregate particles – as in the alkali–silicate reactions (section 6.3). Fine aggregate particles may be harmful for the following reasons:

1. because they are reactive with the embedding medium
2. because they are soft and friable and therefore reduce strength, or weather out
3. because they cause discoloration
4. because they are elongate, and thus create lamination in the product.

The commonest deleterious minerals found in fine aggregate are clay, coal and carbonates.

As described above, one of the reasons for the grading specifications for fine aggregate is to exclude **clay** from the mix. As far as is known, the harmful effect of clay is that it introduces a weak element into the mixture, and by forming a thin coating on the larger particles, prevents the adhesion of either the cement or asphalt to them. It seems that strict observance of the limit of clay content is particularly important in the asphalt used for wearing courses, and for base courses, in road construction; it appears somewhat less necessary for concrete mixes, while for mortars the standard limits are often disregarded, on the grounds that a percentage of clay makes for a smoother finish.

The presence of expanding clays (smectites) is more significant, since their ability to absorb water causes problems in mixing, and subsequent shrinkage of mortars and renderings. This problem has been particularly acute in the use of the Pliocene (Diestian) sands in Belgium and the Netherlands. These are marine glauconitic clays, and it seems likely that it is the alteration of the glauconite to smectite which has been the cause of the problem. A standard specification has thus been established which limits the glauconite to 4%; but it is difficult to apply since there are no easy methods of measuring the glauconite content. In other circumstances, glauconite itself seems to present no problem; it is an important constituent of the greensand deposits of south-east England, which have been extensively exploited for construction purposes over a long period, with no apparent detrimental effects.

Smectite is also found as a weathering product of the ferromagnesian silicates; these of course are primary constituents of basic and ultrabasic igneous rocks. The reactivity of particles of weathered basalt has been a source of concrete failure in the Midland Valley region of Scotland; here there are extensive outcrops of Upper Palaeozoic basalts, which not only form an important source of crushed rock aggregate, but also are a constituent of most of the glacially derived sand deposits. Weathering has deeply affected all these basalts, extending deeply along joints and fissures to beyond the depth of even the largest quarries, so that some weathered particles can be found in every sample of sand from the crusher; the particles in the drift deposits are almost invariably decayed. It follows that concrete made from such aggregate must have doubtful durability.

Small fragments of **coal**, **lignite** or **peat**, while of little importance in mass concrete or in road construction, are of great significance in the material for concrete roof tiles. These tiles have thin sections where they are made to overlap, and the presence of a soft coal particle in this section can cause the tile to leak or crack. Coal can occur as detrital fragments derived from erosion of underlying coal-bearing strata, or as lignitized fragments of plant material indigenous to the depositional environment; because of its low density it is not sorted from the sand by current action, and particles of several millimetres in diameter can be found within the sand. Since also there appears to be no infallible way of removing the coal from the sand, such occurrences can render a whole sand deposit unsuitable for this purpose.

In Britain, many concreting sands are of glacial origin. Unfortunately, in the course of their southward extension, the glaciers rode over many coal-measure outcrops; and few glacial or fluvioglacial sands are entirely devoid of coal fragments (Figure 4.3). Exceptionally, local ice movements passed over younger or older strata, so that by studying the sub-glacial geology, and the known ice movement directions, it is sometimes possible to predict the occurrence of coal-free glacial sands. Figure 4.3 shows the direction of ice movement, and sub-glacial outcrop of coal measures, in the Midland Valley of Scotland. It will be evident that most areas are 'downstream' of coal-measure outcrops, and therefore their sands will contain coal; and that the only areas likely to contain coal-free sand are northern Ayrshire (A), the margins of the central Highlands (B), and a limited area of the Southern Uplands (C).

To a large extent, however, manufacturers of concrete roof tiles tend to prefer to use poorly consolidated ancient sands, such as the Triassic Sherwood sandstone in the Midlands, or the Cretaceous greensand of the south-east – being deposited under arid desert, or shallow-marine conditions respectively, these can be guaranteed to contain no coal. However, it is rarely possible to obtain the correct grading from these latter sands, so that some coal-free Quaternary sand has to be sought to provide the right grain-size distribution.

Bitumen particles in the sand can cause discoloration of the surface

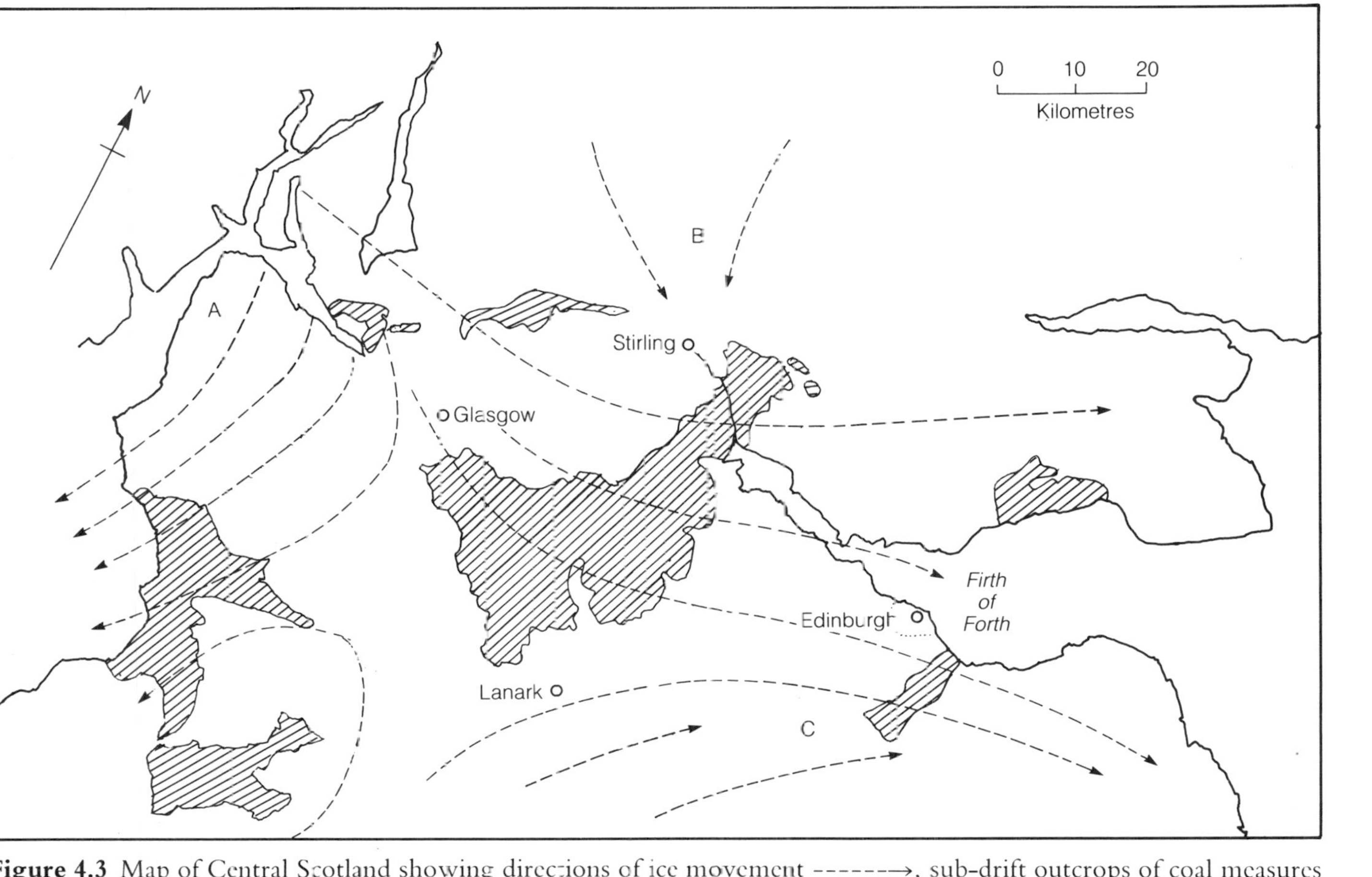

Figure 4.3 Map of Central Scotland showing directions of ice movement ------→, sub-drift outcrops of coal measures (shaded) and possible occurrences of coal-free sand (A, B and C)

of a concrete product, which may be unacceptable in certain uses; and bitumen is a much commoner mineral than is often supposed. For instance, a crushed granite sand from Leicestershire was found from time to time to be unsuitable for concrete block manufacture because of the presence of bitumen; this was traced to petroliferous material associated with some hydrothermal veins in the granite mass, and could be eliminated by selective quarrying.

Carbonate minerals occur in quartzose sands as shells, concretions or thin limestone bands. In many parts of the world carbonate rocks form the only source of aggregate; thus in many Middle Eastern countries surface calcrete is used; in the Mesozoic and Tertiary fold-mountains of the world, crushed limestone, dolomite or marble is the only source; and in many tropical areas the much-used beach 'sands' are entirely carbonate.

Fortunately, the incorporation of relatively pure lime sand into a concrete or asphalt mix does not cause too many problems. Difficulties do arise, however, where limestone is only part of a mixed sand, and when the limestone is dolomitized.

One of the problems of a sand containing some limestone is that of differential shrinkage. All limestones have some degree of porosity, and it can be up to 28%, while quartz particles have virtually no porosity. On incorporation into a wet mix of concrete the limestone particles slowly absorb water, so that after emplacement the concrete gradually shrinks. Moreover, if the distribution of limestone particles is not uniform, some areas shrink more than others, so that cracks develop, which can result in the total disintegration of the concrete.

Chemically, cement and pure limestone are non-reactive, and so it is perfectly possible to make a durable concrete using wholly limestone aggregate. There is, however, interaction between high-magnesium calcite or dolomite and the cement matrix. Thin sections of concrete show distinct reaction rims around dolomite particles which must introduce internal stresses. The problem has been found to be particularly acute in Czechoslovakia, which is almost wholly dependent for its aggregates on the large dolomite masses of the Carpathian mountains. It has been shown that loss of strength in a concrete can be related particularly to the fine fraction ($<63\,\mu m$) of the aggregate; marked loss of strength occurs when the fraction $<63\,\mu m$ exceeds 20%. The effect is more marked with higher strength concrete, which shows a steady decline in strength from 5% upwards (Figure 4.4).

While the above sections describe most of the minerals found in fine aggregate which consistently result in poor quality concrete, there are others which may in certain circumstances be troublesome.

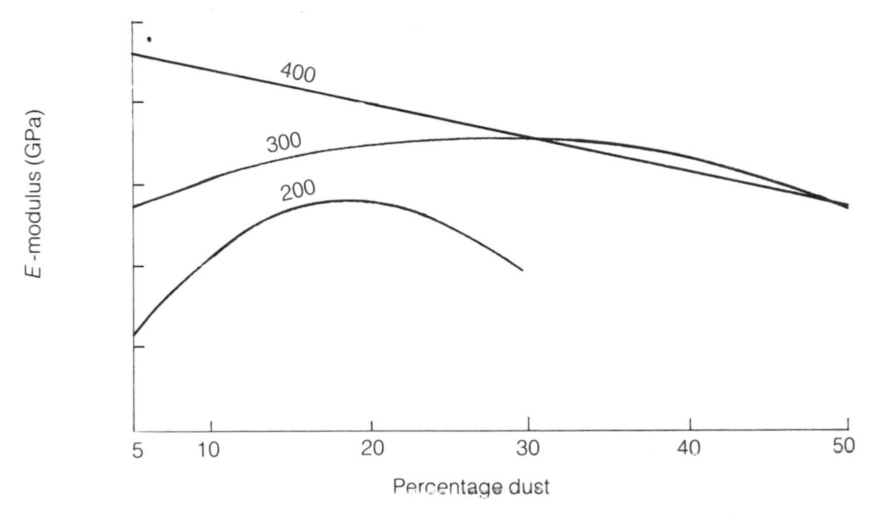

Figure 4.4 Graph showing loss of strength in concrete with increase of fraction <63 μm of dolomite fine aggregate. After Rigan and Žabka (1976, Figure 6).

Platy minerals, of which the commonest is mica, tend to align themselves in the concrete if it is extruded or vibrated, creating a laminar structure which introduces directional weakness. Such mica can be detrital in origin, where a sand is derived from, for instance, an area of schistose rocks. Many sands used for fine aggregate in Norway, for instance, have this mineral in abundance. Where the erosion source of the sand is a terrain of slaty rocks, flat, sand-sized particles of these may form a substantial part of the sand. For instance, the sands of south Lanarkshire (Area C in Figure 4.3), while free from coal fragments, are rendered unsuitable for some concrete production because they consist very largely of such platy fragments.

Friable particles, of which coal (see above) is perhaps the commonest example, can be of various other mineralogies. Particles of soft, poorly cemented sandstone, for instance, are found in many Quaternary sands where the outcrop of such sandstones is a local erosion source.

The problem of friable particles is particularly acute in areas where colluvial sediments form the main source of sand. Such deposits are formed by the rapid degradation which occurs in mountainous regions; they are often deposited quite close to their erosion source, and in the short distance of transport there has been no opportunity for attrition processes to have removed the softer particles. A good

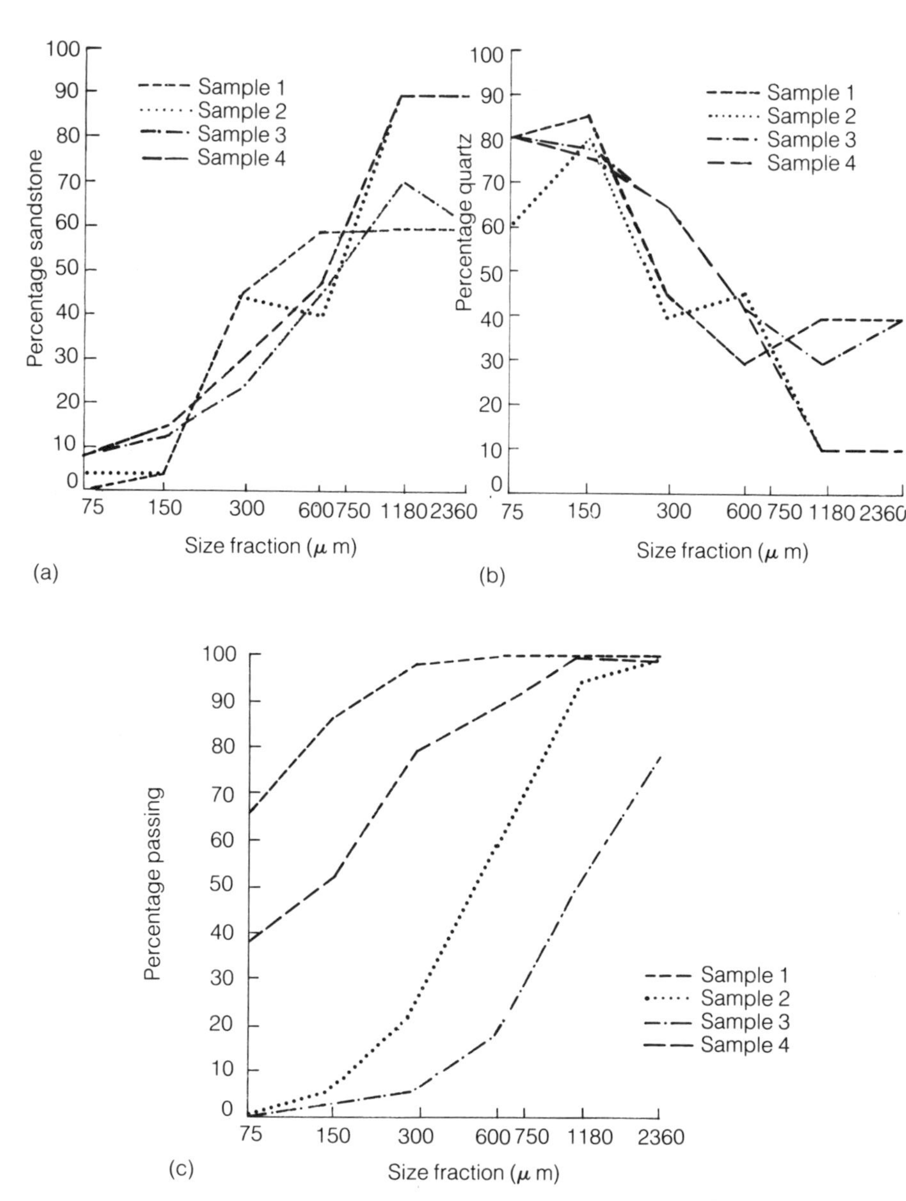

Figure 4.5 Mineralogical and grain-size analyses, fluvioglacial sands north of Stirling, Scotland: (a) percentage of sandstone grains in each size fraction; (b) percentage of quartz grains in each size fraction; (c) grain-size distribution of sand samples; (d) percentages of sandstone, quartz, metamorphic rocks and feldspar grains in each size fraction of a sand from a more northerly deposit; (e) grain-size distribution of (d).

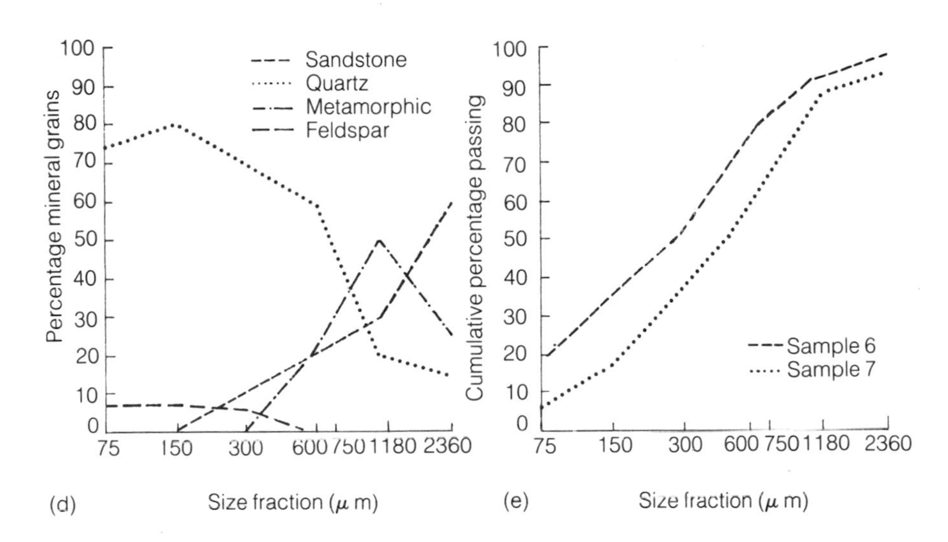

example is in the Madrid region of central Spain, which depends for its construction sand on a series of Neogene and Quaternary sands, which are directly derived from the wasting of the granitic Sierra de Guadarrama, immediately to the north. In consequence of this proximity, these sands contain large quantities of feldspar particles, producing a concrete of low strength and doubtful durability. Sands with higher quartz contents are found where there has been secondary reworking of such sands, as has occurred in some more recent river deposits, and these are much sought after.

There is a manifest relation between mineralogy and grain size. The size of particles found in a sand deposit is dependent upon

1. the original grain size as weathered from the rock source
2. the amount of attrition between source and area of deposition, and
3. the sorting during transport and deposition.

Since each mineral has a different original grain size, a different hardness, and a different specific gravity, a homogeneous distribution of minerals throughout the size gradings is not to be expected. This can be illustrated by reference to samples from a fluvioglacial sand in an area north of Stirling, Scotland, on the margins of the central Highlands (Figure 4.5). The bedrock is Old Red Sandstone, and the sands contain an abundance of friable sandstone particles derived from the bedrock, accompanied by quartz grains from more distant sources. An analysis of each grain-size fraction (Figure 4.5(a) and (b)) shows an antipathetic relationship of the two major constituents, with sandstone predominating in the coarse fraction, and quartz in the fine. Study of the grading curves of the same samples (Figure 4.5(c)) shows that the sands have different depositional histories, samples 1 and 4 differing markedly from 2 and 3, but this does not affect the quartz/sandstone relationship – suggesting that source and transport are more important determinants of this than depositional agencies. A related sand deposit a few kilometres to the north (Figure 4.5(d) and (e)), nearer the metamorphic terrain of the Highlands, contains schist fragments in the coarser grades, while feldspar appears in the fine fraction – its small size perhaps reflecting the long distance it has travelled. The grain-size distribution shows that yet another depositional environment is here reflected.

The message from such studies is abundantly clear – that the mineralogy of a sand is closely linked with its granulometry; and that two sands from the same deposit might have a different mineralogy because they have a different size distribution. A further complication is introduced by the processing of a sand. The process of washing or hydrosizing, while mainly achieving a sorting by size, can also effect a mineralogical sorting. This is because minerals, by reason of their differing specific gravities, and surface effects, behave differently hydrodynamically in the washing process. Thus washing can for instance increase the quantity of basalt particles in a sand, because their higher specific gravity makes them behave as though they were comparatively larger than quartz. And a sand produced by washing can have a different mineralogy, for the same reasons, from one produced by dry-sieving.

The same strictures apply to sand produced by crushing rock; we have seen above (section 3.3) that different crushing processes produce a different granulometry. The size of broken grain produced from each mineral will also be different; so that unless the rock is monomineralic, the crusher product will vary in its mineralogy in each size fraction.

It follows from the above that, except where there are known wholly monomineralic sands, routine mineralogical analysis of a sand is essential. Under normal circumstances, all the necessary identifications can be carried out by examining the sand under a low-power binocular microscope (Figure 4.6). Distinguishing features of colour, texture, cleavage, etc., which will be familiar to any geologist, can best be seen if the sand is immersed in water, and quantification can be achieved by counting individual grains, or by a point-counting technique. Limestone grains can be identified by

Figure 4.6 Fine aggregate under low-power binocular microscope.

moistening with dilute acid, and dolomite distinguished by strong acid.

A useful technique in the field of concrete production is to cut and polish a section of the concrete product, and to examine this under a binocular microscope (Figure 4.7). This is particularly useful as it enables one to see the interior of grains, thus allowing more precise identification, and revealing the presence of weathering, friable textures, etc. in individual minerals. It is helpful in distinguishing, for instance, feldspars from quartz, not only because it reveals the cleavage, but also because it enables staining techniques to be carried out on the feldspars. It also shows the behaviour of minerals in the product; lamination effects due to the presence of platy minerals are clearly revealed, and reaction rims around active minerals identified. Sometimes it is necessary to cut thin sections, in order to identify an unusual mineral, for instance; these can as easily be cut from the concrete product, as from the embedding materials usually used in petrography.

Other sedimentological techniques can be called in to assist; heavy minerals can be separated by sedimentation in bromoform; iron minerals by magnetic separation.

The emphasis in concrete and asphalt technology, in research and in standard specifications, has consistently been on grain-size analysis. The importance of mineralogy has been less considered. This is partly due to the fact that the western world is mostly in regions whose recent geological history has provided abundant supplies of quartz-rich sand. As these supplies become exhausted, and as modern technology extends to areas whose geology is not so

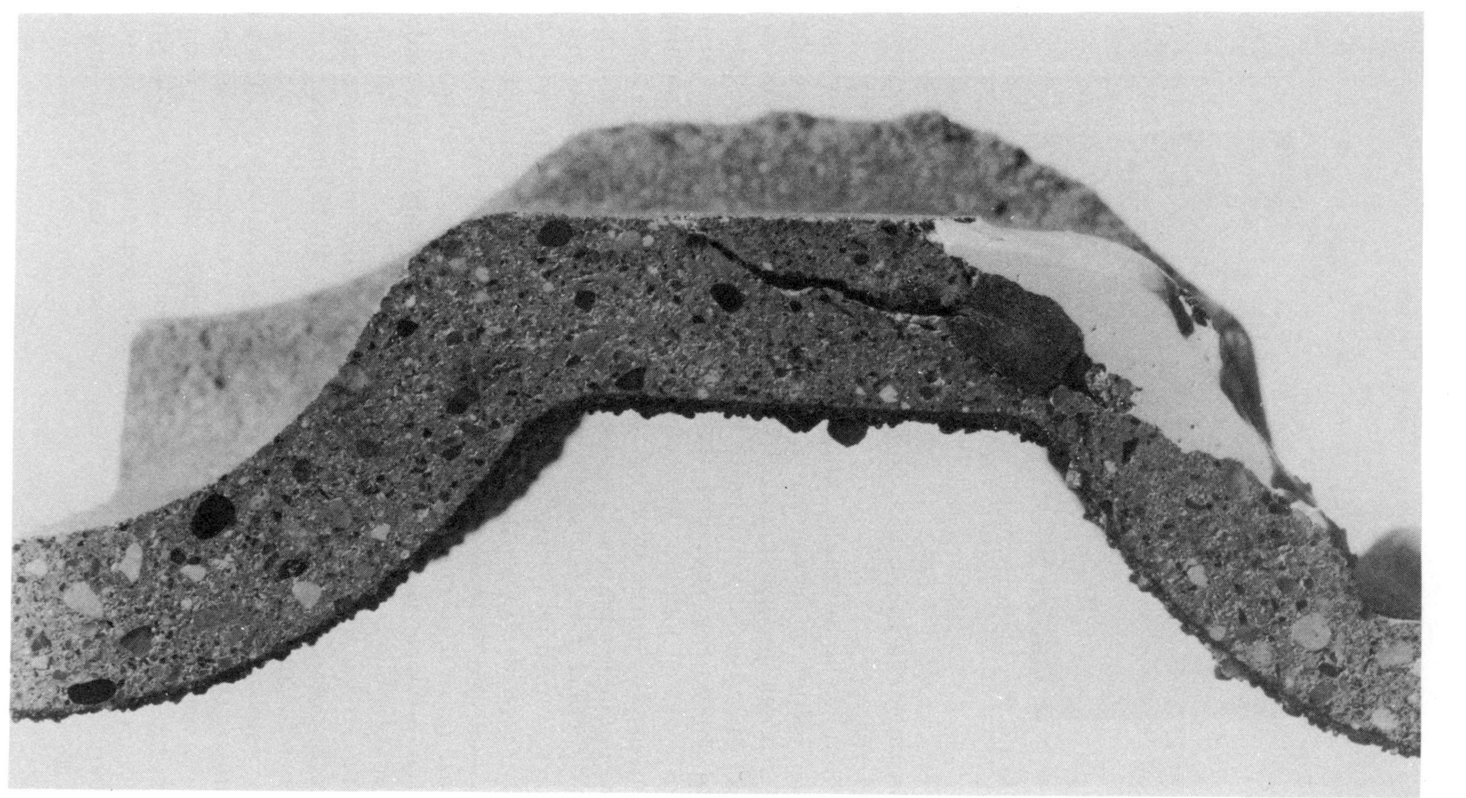

Figure 4.7 Polished section of concrete; note large expansive particle which has caused cracking.

favourable, supplies of sand need to be sought whose mineralogy is much more variable. The disasters consequent upon the alkali–silicate reaction in concrete, and the problems of concrete construction in the Middle East, have, in particular, alerted the civil engineer to the significance of mineralogy, and it is certain that in future the geologist will be required increasingly to contribute his or her skill to an understanding of the mineralogy of sand.

4.4 SOURCES OF FINE AGGREGATE

A small amount of fine aggregate is produced as a result of crushing rock; but this is usually incidental to the production of coarse aggregate. The manufacture of fine aggregate from rock, which would involve an expensive milling operation, is rarely undertaken, except in areas where natural fine aggregate is extremely scarce. The majority of fine aggregate is thus derived from natural sources.

Fine aggregate occurs along with coarse in the various geological situations described in section 3.4 – colluvial, alluvial, marine or glacigenic; but there are differences in distribution of these.

Colluvial sands are comparatively rare, since the process of formation usually results in gravel rather than sand; but there are some exceptions. Granitic rocks, whose original mineral size approximates to that of sand, may be broken down by chemical weathering – the feldspar and mica decay, leaving a loose mass of quartz crystals, which are then accumulated in detritus deposits near their source. Since these deposits occur in areas where chemical weathering is intense, they sometimes provide a source of sand where other sources are scarce – as for example in central Spain, and in the Transvaal.

Most alluvial deposits consist of a mixture of gravel and sand, so that coarse and fine aggregate are produced together. However, in mountainous regions, where the rivers are normally torrential, the sand may have been swept away, and the material dug from the river bed may be coarse aggregate only. This happens for instance in the south of France, in Spain, and in Italy, so that there is in these areas often a notable deficiency in the supply of sand, which is sometimes supplemented by the products of crushing oversize material.

Marine deposits, of shingle beaches and sea-bed type, also contain a proportion of sand. But it is characteristic of the marine environment that sorting of sand from the pebbles occurs, so that there are many places where sand occurs on its own – as sandy beaches along the shore, or as offshore sandbanks. These are emplaced by the waves and tides, and the sand grains are in constant motion; in consequence attrition and rounding of the sandgrains occurs very rapidly, and most beach sands are too fine for many uses (section 4.5).

Glacigenic deposits, particularly those of fluvioglacial origin, usually contain both gravel and sand. Debris being transported away from the ice-front by meltwater streams is deposited very rapidly, so that sorting is imperfect; thus near the former ice front an esker will consist of an irregular mixture of all sizes of sand and pebbles, and as this is traced downstream, the percentage of sand will increase, so that distal parts will consist entirely of sand and silt – and such deposits are often an important source of fine aggregate. Sometimes the meltwater, and its sediment load, terminates in a glacial lake, which results in a better sorted deposit; but such deposits are mostly too fine for industrial uses.

A sediment-transporting agent which is capable of moving sand, but not gravel, is **wind**, and huge accumulations of sand dunes occur in desert areas, and along sea-shores. However, the attrition processes in wind transport are as effective as that of the sea, and so such sands are usually extremely fine grained; also, normal wind velocities are only adequate to transport the finer particles. Moreover, in the case of coastal dunes, their role as coast-protection agencies is well appreciated, and there would be the strongest environmental objections to their being exploited.

During the late stages of the Pleistocene glaciation, when the ice-cap was retreating, and the lands to the south were bare tundra, there were strong and regular winds blowing outwards from the ice-bound regions. These were able to pick up loose fine-grained material from the abundant debris brought down by the ice. The finest, clay-grade material was carried far to the south, and deposited as a thick sheet of **loess**; but the slightly coarser sandy material was deposited first, and remnants of a sheet of such wind-blown sand are found in a belt extending from East Anglia across into Holland, north Germany and Poland. It is an indication of the strength of these Pleistocene winds that some of these sands are coarse enough to have industrial uses.

Those of us who live in northern and temperate latitudes are accustomed to fine aggregate being composed of grains of quartz, with only subsidiary quantities of other minerals. This is because the terrain from which the sands were derived is dominantly igneous and metamorphic, and therefore contains much silica: that terrain has been eroded in climatic conditions in which mechanical, rather than chemical weathering has prevailed – a situation which favours the preservation of quartz: and recent glaciation has spread this quartz-rich debris very far south. There are many parts of the world where these conditions. do not obtain. In many of the young folded mountain areas of the world – as for instance in the eastern Mediterranean, and much of the Middle East – the terrain of origin consists very largely of limestone, so that all the derived sands, alluvial, marine or

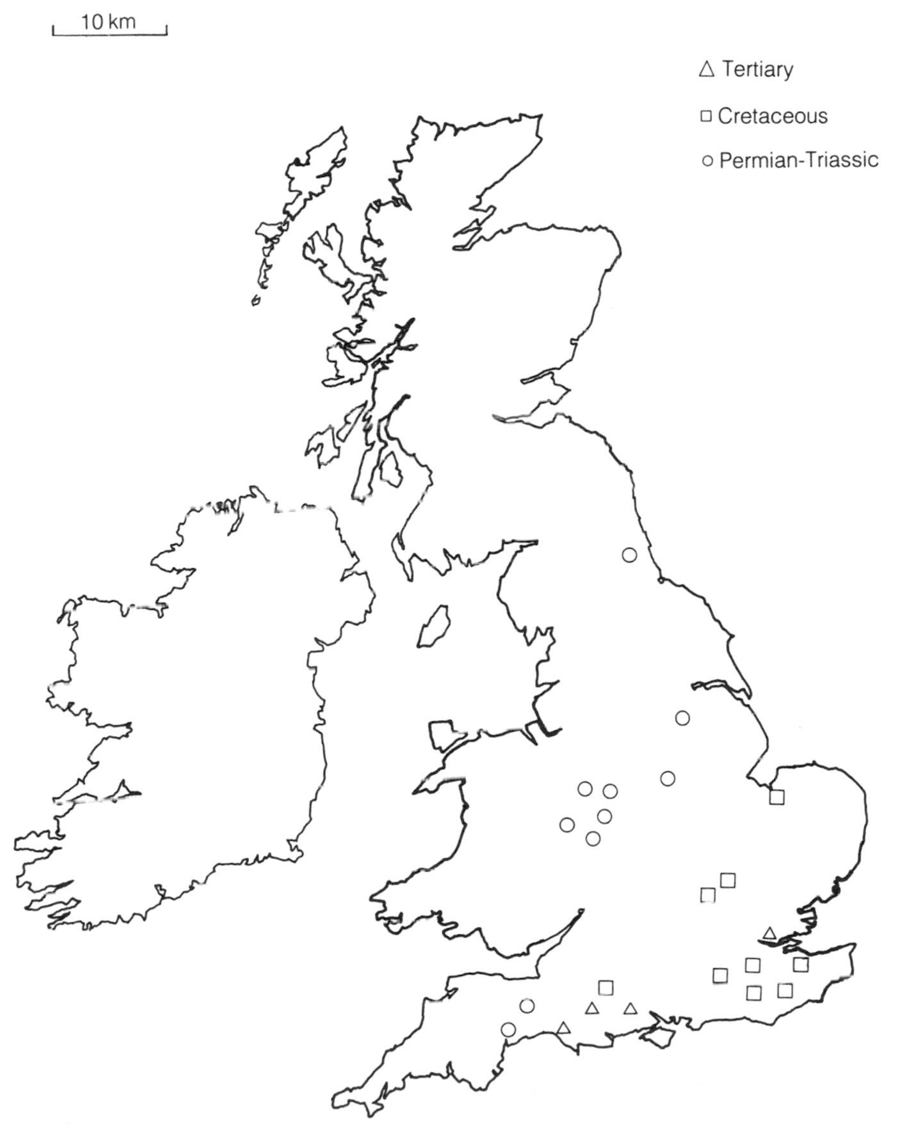

Figure 4.8 Regions of Britain which produce fine aggregate from pre-Pleistocene sources.

whatever, consist of grains of calcite. In areas where there has been recent vulcanicity – as, for example, in Iceland – all the grains of sand are composed of basalt. And in those regions whose recent history has been the formation of coral reefs, grains of broken coral constitute the sand formations.

Sands which are sufficiently unconsolidated to dig, and to convert to fine aggregate with minimal processing, are found in many older geological formations. In Britain, the oldest such rocks are in the Sherwood sandstone of the English Midland area (Figure 4.8). These were deposited in the inland seas of the then-desert area, and have a strong red colour from the oxidized iron of the desert environment. The iron is not apparently deleterious, but imparts a red colour to a mortar or cement mix.

Higher in the geological sequence are the sands of the Cretaceous – the so-called greensand. These are marine in origin, having been deposited as sandbanks in the shallow basins of the extensive Cretaceous sea. They contain substantial quantities of the iron mineral glauconite ($K(Fe^{2+}, Mg, Al)_2Si_4O_{10}(OH)_{10}$), which, although it is known for the facility with which it converts to smectite, does not appear to have been a disadvantage in its extensive use for concrete in the south-east of England. While locally there is cementation by silica, limonite and calcite (giving rise to some stone sufficiently hard to be used for rural building), most of the greensands are loose and unconsolidated. Similar deposits of this age are found in Belgium, around the Paris basin, and southwards into Aquitaine and Provence – but pass into a limestone facies when approaching the Alpine and Pyrenean mountain zone.

Although sands are abundant in the Tertiary formations of Britain, Belgium and northern France, they are in general much finer than those of the Cretaceous, and have thus a limited utility. They also contain glauconite, some of it derived by erosion of the Cretaceous greensand, and this has been known to be reactive in concrete. The warm seas in which they were deposited supported an abundant molluscan fauna, and thus many of these Tertiary sands are shelly – a further factor in diminishing their utility. The Tertiary rocks were deposited in a series of shallow marine basins; in the centres of these basins the rocks are predominantly clay and silt, and any sand deposits are very fine grained. Towards the massifs surrounding the basins the sands become coarser; and these are useful for fine aggregate on the margins of the Armorican massif (west of England and Brittany) in the west, and the Brabant massif (southern Belgium) in the north.

4.5 GRAIN-SIZE DISTRIBUTION OF FINE AGGREGATES

The various uses of fine aggregate demand different characteristics – the chief variable being grain-size distribution. There are standard specifications for each of the major uses in most countries of the world, and these vary from country to country. The British Standard Specifications are quite typical and so can be used as an example.

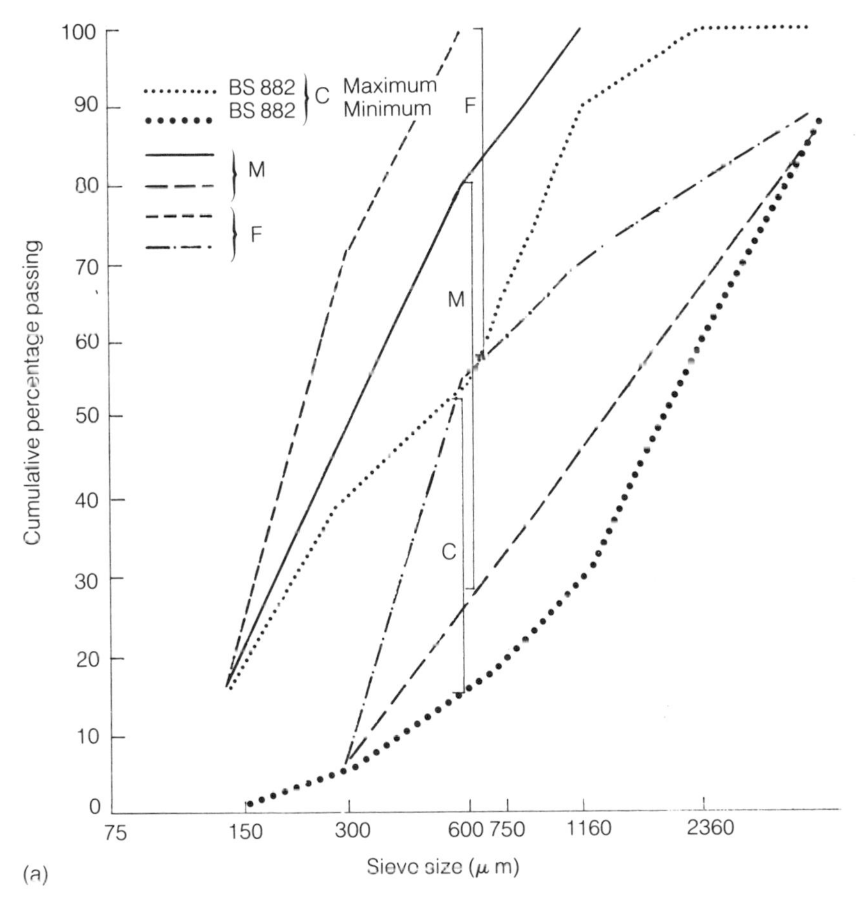

Figure 4.9 British Standard Specifications for: (a) concrete sand (BS 882); (b) asphalt sand (BS 594); and (c) sand for mortars (BS 1200). Size distribution of examples of sand from different geological environments: (d) sea-dredged aggregate; (e) coastal deposit; (f) Pleistocene blown sand; (g) fluvioglacial deposit; and (h) alluvial sand.

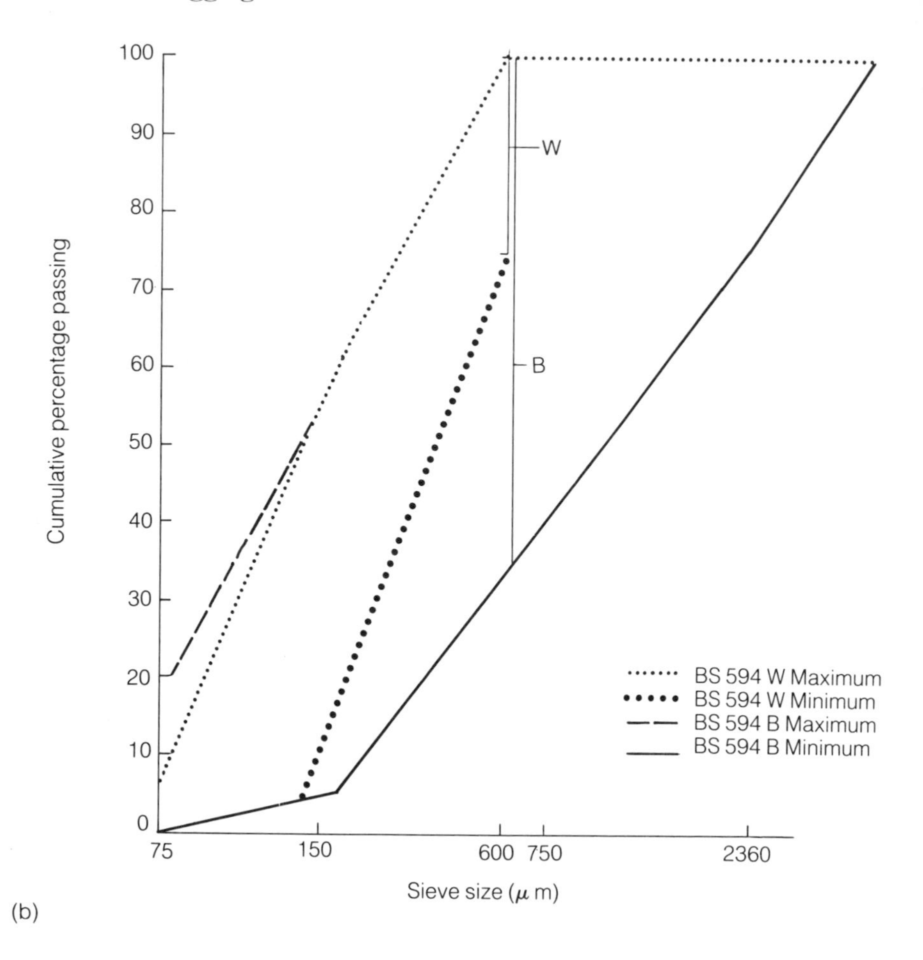

(b)

The British Standard Specifications for fine aggregate for use in concrete, asphalt for roads, and mortars, are illustrated graphically in Figure 4.9. The specification for concrete distinguishes three grades – C, M and F. Perhaps the most significant differences lie in the 'passing 600 μm' region, whose limits are 15–54% for C (coarse), 25–80% for M (medium), and 55–100% for F (fine). Apart from specifying that only the C and M grades can be used for heavy duty concrete, there is no indication in the standard specification that these grades have specific uses; but there is a general understanding that the coarser grades make stronger concrete. Sand for the asphalt used on roads (Figure 4.9(b)) is generally finer, although the specification for the top 'wearing course' is much narrower than for that of the 'base course'. For asphalt great importance is attached to the content of

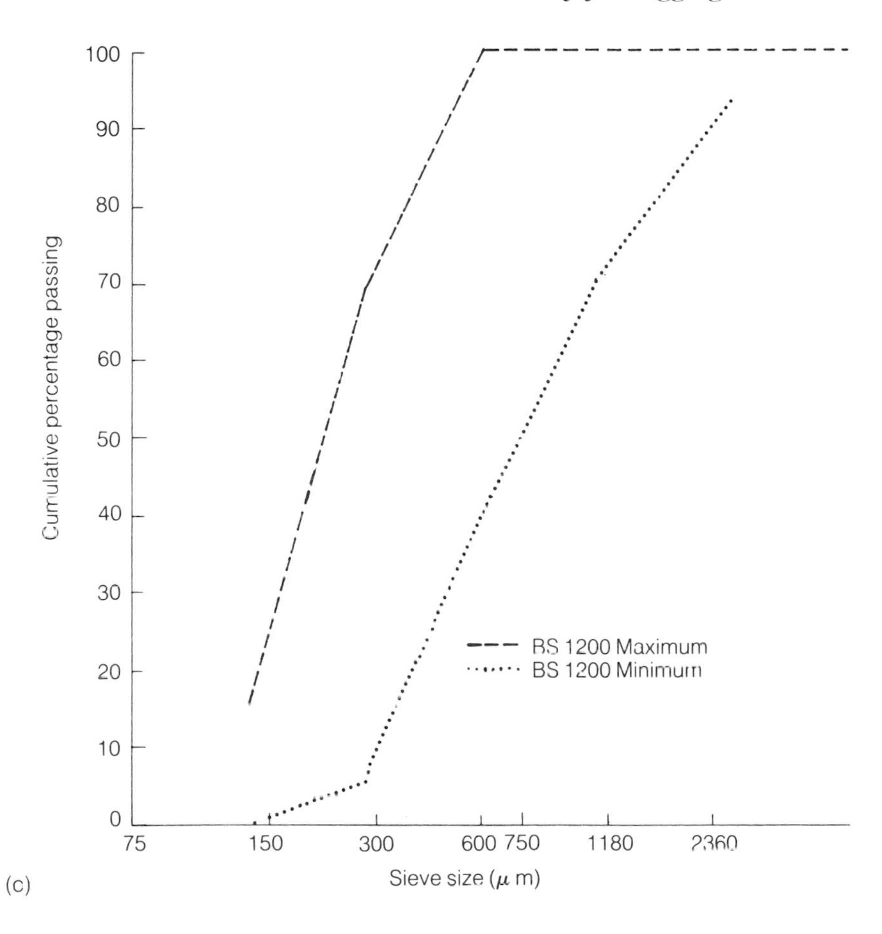

fines (i.e. <75 μm), which may not be more than 5% for wearing course and 18 % for base course. The specification for 'general purpose mortars' is typical of a number of specifications for use in mortars, renderings, and external and internal finishes. These specifications are for finer sand than for concreting, and it is generally agreed that a good percentage in the 150 μm to 300 μm range is important in ensuring workability and a smooth finish. It must be said, however, that adherence in the industry to this last group of standards is minimal, and many 'building sands' are sold, and successfully used, which lie outside the specification limits.

It is clearly a commercial advantage if the sand extracted from a pit can be sold 'as won', or with a minimal amount of processing. With simple screening (to remove oversize) and washing (to remove

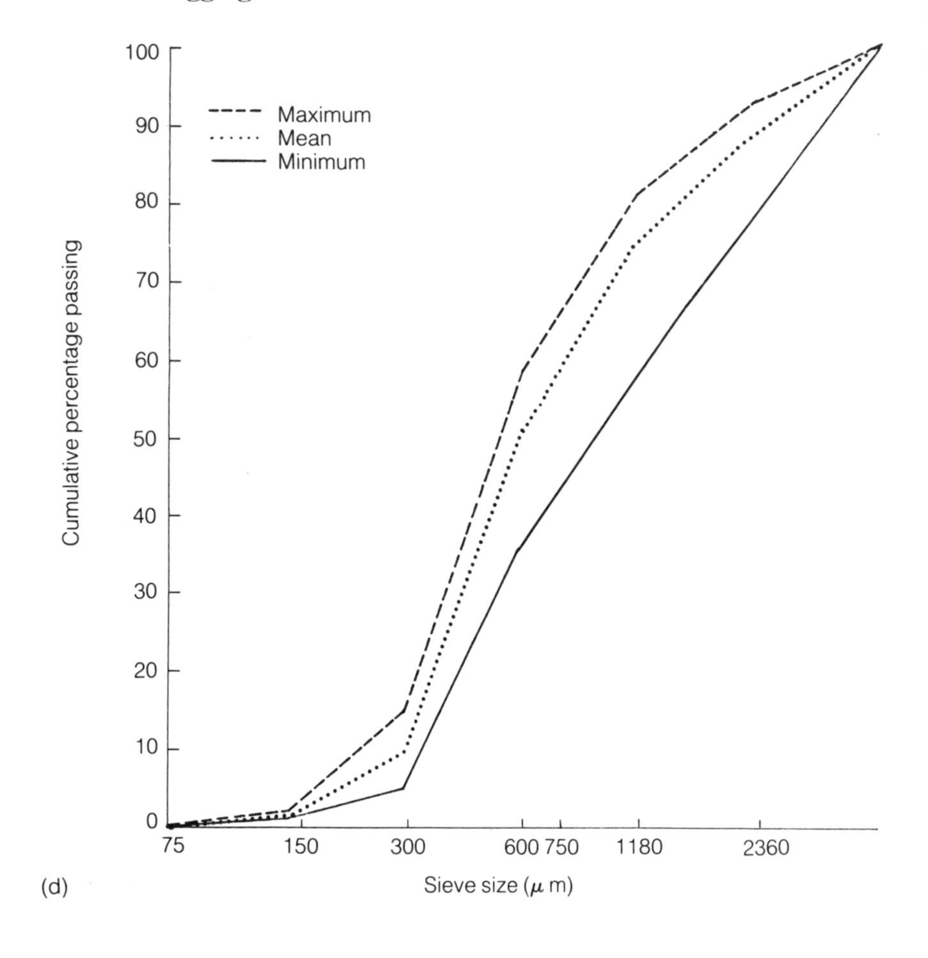

(d)

undersize silt or clay), many natural sands can be made to conform to the standard specifications. This is not entirely surprising, since the standard specifications are based not on a scientific appraisal of the best or the ideal material, but on the experience of industry as to which currently available sands are known to give a satisfactory product.

Figure 4.9(d) to (h) shows the size distribution curves of some naturally occurring sands. An example of the sand extracted from the sea-bed dredged aggregate is shown in Figure 4.9(d); it will be seen that the range of variation is quite small, and that all samples lie within the limits of concrete sand C (Figure 4.9(a)); while it is clearly 'out-of-specification' for asphalt (Figure 4.9(b)), and only just within that for mortars (Figure 4.9(c)).

Figure 4.9(e) is typical of the sand extracted from shingle ridges,

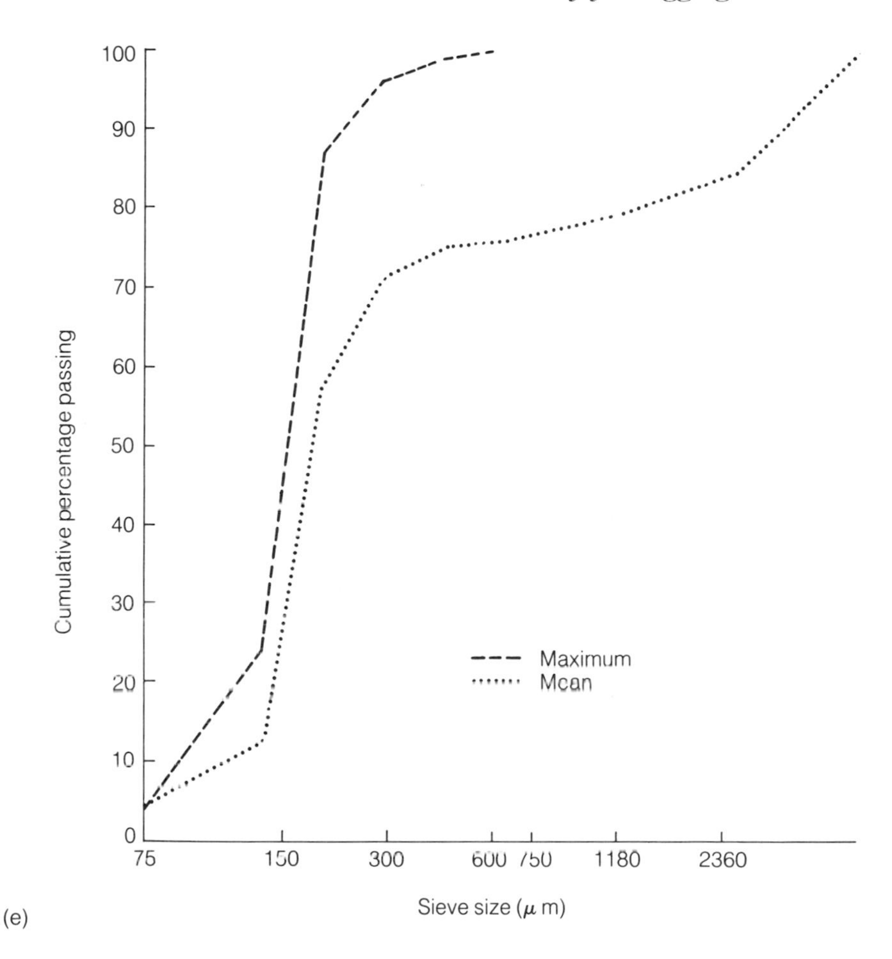

(e)

and from many modern beaches. It is notably deficient in the coarser grains (>300 μm), and does not lie within any specification. Such a sand could be processed to yield some building sand, by removing the finer grades, but this would clearly be very wasteful. The Pleistocene blown sands, illustrated in Figure 4.9(f), show a more regular grading, and although 'as won', some of the samples lie outside specification for asphalt wearing course, they could be brought within it by removing at most 10% of the 'fines'.

In great contrast is the fluvioglacial deposit (Figure 4.9(g)), which shows the size distribution from an esker deposit in central Scotland. The range of variation is enormous, from the fine silty clay indicated by the left-hand curve, to the coarse clean sand of the curve on the right. It is typical of such deposits that the various types of sand occur in such a way that selective extraction is not possible; but

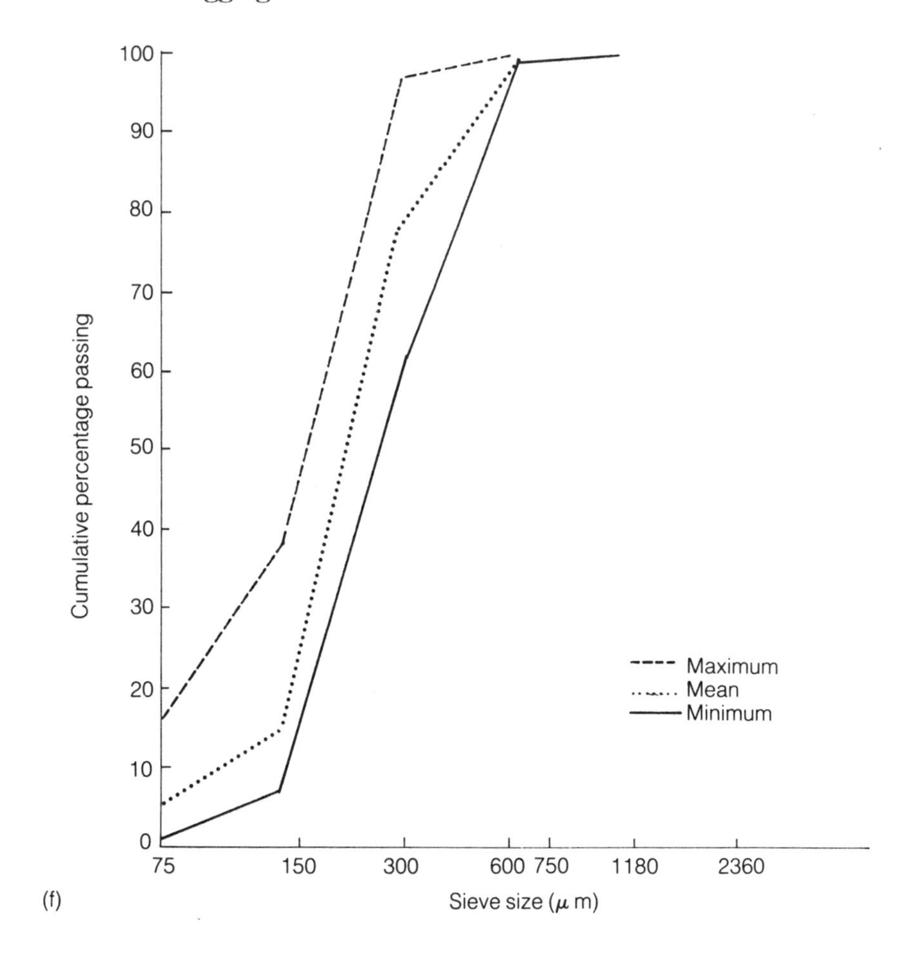

quarrying this deposit will give a raw material containing a wide range of sizes; and careful processing will yield a variety of sands which can be made to conform with the various specifications.

Finally, a typical river-terrace deposit is shown (Figure 4.9(h)). The narrow range of variation is characteristic of the sorting capacity of river transport. This particular example fits neatly within the M (medium) concreting sand specification; and other examples can be found which lie within the C and F limits. For obvious reasons, these sands are much favoured for the production of concrete.

4.6 SHAPE AND TEXTURE OF SAND GRAINS

Although these factors are known to be highly variable, no systematic study of their effects upon the products has been made. It

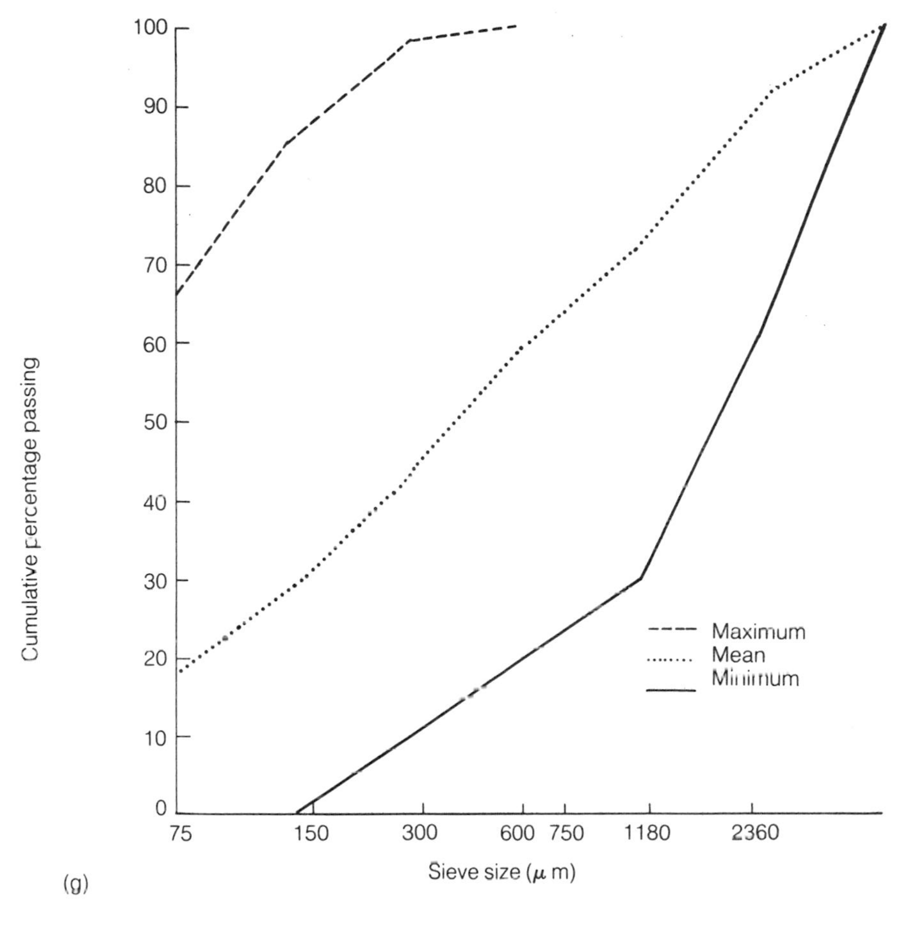

(g)

is known that angular, rough-textured grains impart some strength to concrete – presumably because the surface gives better adhesion for the cement. At the same time this decreases the workability of the concrete, and tends to give a rougher finish. In concrete products which are extruded, such as concrete tiles and pipes, the more angular fragments may give an excessive abrasiveness to the mix, which can be costly in machine wear.

Sand grains in fluvioglacial deposits are the most angular and rough textured. Alluvial grains are more rounded, with a smoother, but often pitted, surface. Marine transported sands are usually well rounded, with a smooth surface, while those transported by wind are often nearly spherical, with a very smooth, polished surface. Thus, as with size distribution, there is a need to match the characteristics of the deposit with its ultimate use.

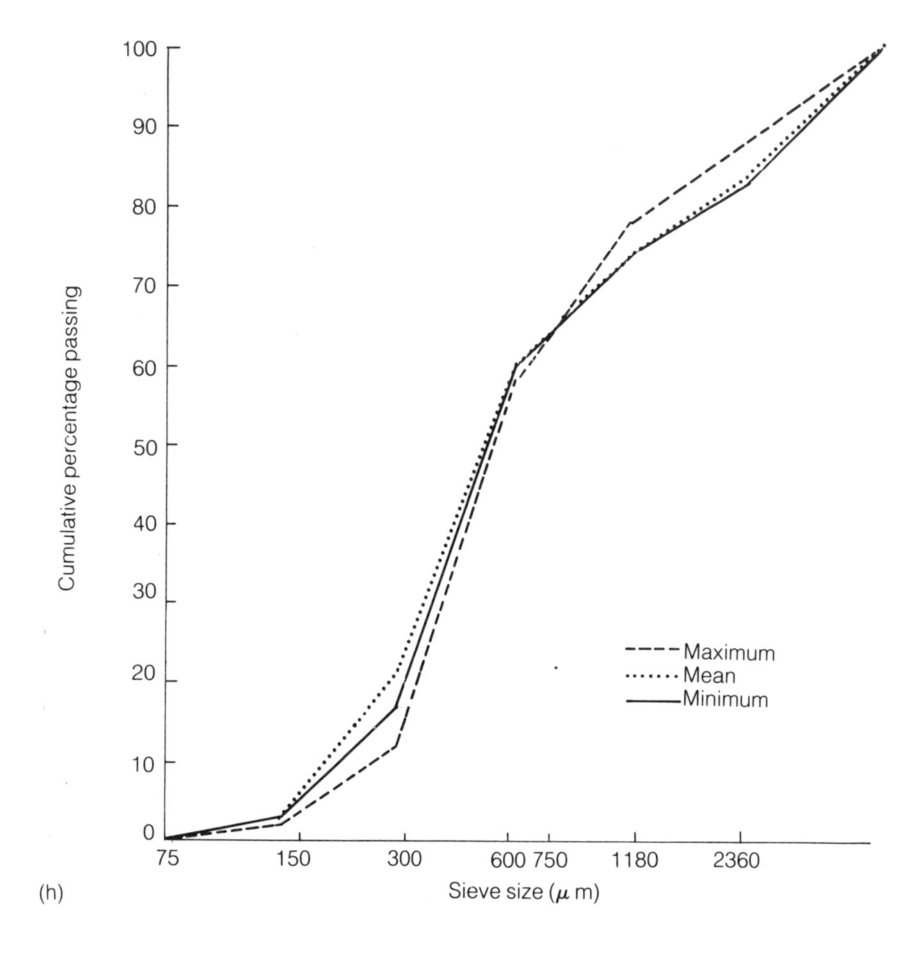

Fine aggregate derived from crushed rock will obviously be much more angular than any natural sand; apart from this the shape, texture, and size distribution will be to some extent controlled by the original mineralogy of the rock from which it is derived. Thus crushed gneiss will yield a sand whose particles vary in size, shape and texture, while a crushed limestone will give uniform, equidimensional particles.

5

Structural clay products

5.1 INTRODUCTION

The art of baking clay to produce a brick, tile or pipe is a very ancient one indeed; fragments of burnt brick have been found from before 1200 BC in Egypt, and even earlier from Mesopotamia and India. The usual building material in the ancient civilisations of the warmer climates was, however, a sun-dried brick, usually containing animal dung or chopped straw. This product served to build most of the pyramids of Egypt, the walls of Jericho, and a large part of the Great Wall of China. It is the Romans who must be credited with the development of the burnt brick; they brought its production to a very high standard by weathering the clay, by fine grinding, and by controlled burning. The durable bricks, reused, served as a building material through the post-Roman period, and it is not until the twelth century that we have evidence of indigenous brick production in the northern civilisations. Since then, however, brick and tile production has been ubiquitous. In the eighteenth century every village, every estate, had its own brickworks, using surface clay dug by hand from the immediate locality, and processed and fired by methods which had changed little since Roman times. The industrial revolution brought a demand for large-scale production, and mechanization of the process began. At the same time brickworks became larger, and the concentration of the industry into a small number of large units began – a trend which has continued up to the present day. Such large production units dig deeply into unweathered clay; and it is the problem generated by the industry's demand for large-scale clay supply that the geologist is called upon to solve.

5.2 THE BRICKMAKING PROCESS

The procedures in making a brick can be described under five headings – clay extraction, clay processing, brick forming, brick drying and brick firing. The geology and mineralogy of the clay has important consequences at each of these stages.

Extraction of clay is almost universally by open-pit methods, since the low value of the product cannot absorb the high costs of deep mining. Most brickclays are stratified sedimentary deposits; it is rare for such deposits to be completely uniform – usually each layer within the sequence shows some degree of variation. Before the days of mechanization, selection of individual clay seams was possible; but modern clay digging can only exercise a limited selection, and it becomes more important to ensure that the method of extraction produces a uniform and consistent mixture. Each method of extraction has its own characteristics, advantages and disadvantages (Figure 5.1).

The **shale planer** or **multi-bucket excavator** consists of a series of toothed grabs arranged along a continuous chain; the chain of grabs is placed against the face, generally at an angle of about 45°, and the grabs make a continuous cut, discharging the material at the top of the face. The method is particularly useful where the deposit is reasonably homogeneous, as it ensures a constant and even mix, but any selection of seams is virtually impossible. Not least among its advantages is that it leaves a clean face, making the work of the geologist much simpler. This method has lost favour in Britain, because of problems in maintenance of the machine, but it is still used extensively in continental Europe and the United States.

The **dragline** consists of a grab suspended by cable from a long boom; it operates from the top of the face and the bucket is filled by dragging it from the base of the face upwards. The adequacy of the mix is very dependent upon the skill of the operator; often it is necessary to create a pile of broken clay at the base and dig from that – even so it is difficult to ensure a constant mixture. A limited amount of seam selection is possible.

The use of a **tractor and scraper** involves the operator making successive passes across the pit, and scraping a uniform layer of clay with each pass. The effect is that of a very low angle face, and if the dip of the strata is low, the operation is extremely difficult to control. It is, however, particularly useful in the building of a stockpile since the machine can take a cut, then deposit its load, in a continuous run.

Fixed-arm excavators work from the bottom of the face; they are limited in the height of face possible, but close control of mixing and seam selection is possible. The back-acting excavator has similar limitations, but can work easily at the foot of, or on a bench – in the latter situation very precise clay selection is possible.

Clay is sometimes taken directly from the quarry to the processing plant, but more frequently there is an intermediate stage in the

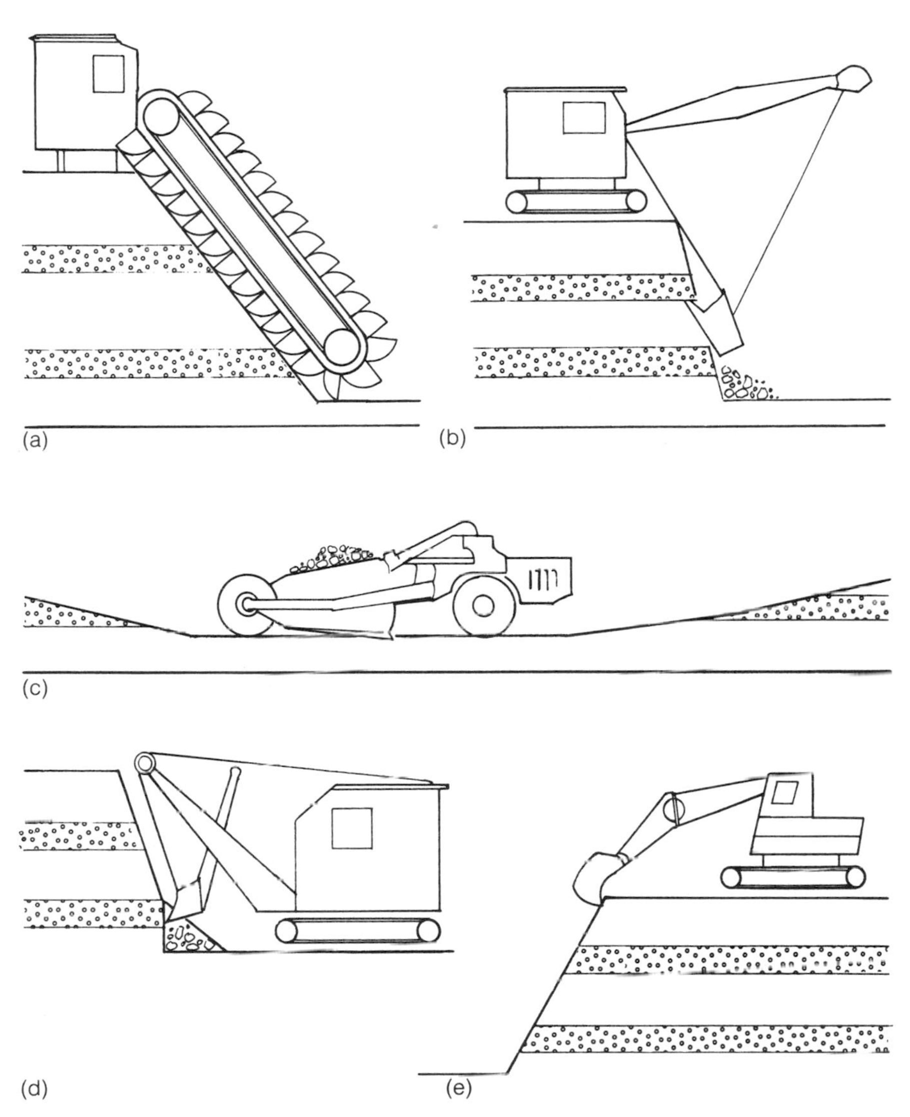

Figure 5.1 Clay extraction methods in relation to strata: (a) shale planer or multi-bucket excavator; (b) dragline; (c) tractor and scraper; (d) fixed-arm excavator; (e) back-acting excavator.

formation of a stockpile. This is a heap of clay, generally about ten to twenty metres wide, and five to ten metres high, built up from the base in successive layers; it is left for several months, generally over a winter period, and then redug by a fixed-arm excavator. The advantages are various. Exposure to the weather, particularly over the winter, breaks up the structure of the clay, making subsequent grinding easier. The whole mass of the stockpile acquires a more or less uniform moisture content, so that this can be more readily controlled. If the clay contains pyrite, this mineral breaks down to sulphates which are at least partially washed out of the stockpile. But, most important, it provides a means of thorough mixing of the clay, and a close control of the clay mixture – the geologist, by supervising the building of the stockpile in the first place, and then by taking regular monitoring vertical sections through the stockpile, can ensure that the mixture going into the plant has the desired qualities for the product.

On entering the plant, the clay is ground by steel rollers; at this stage any water necessary is added, and also any grog, fuel or other additives can be put in. It is important to realize that grinding is rarely carried, for brick production, to a point where full comminution is achieved, so that even after the grinding process the clay is quite heterogeneous. For more specialized products, such as wall tiles, more complete grinding may be achieved by milling in a ball- or pebble-mill.

The next stage is to shape the brick; this can be achieved in a variety of ways. Traditionally it was done by hand; the moulder took a ball of clay, and threw this into a sanded mould; he then cut off the top with a knife, and inverted the mould on to a palette, on which it was then transferred to the next stage. This process produced a characteristic wrinkling of the face of the brick – a feature much prized by architects. In consequence, machinery has now been developed which reproduces the throwing process; and **simulated handmade bricks** are now in large-scale production.

Other machine moulding methods are available to make different kinds of brick, and depend on the water content of the clay. **Dry-pressed bricks**, where moisture content is less than 5 %, are forced into steel moulds under pressure, whereas **soft-mud bricks** and **slop-moulded bricks** (where water content can be up to 20%) can be squeezed or poured into the mould. To assist demoulding, dry-pressed bricks require the mould to be oiled; for soft-mud bricks, sand is the usual agent, whereas slop-moulded bricks require no such agent. Each process produces different surface textures and different internal structures.

Finally, bricks can be formed by **extrusion**. The ground clay is passed through a vacuum to remove air bubbles, then forced through a rectangular die, which extrudes a column of clay. This is then cut vertically by a series of wires. Surface coatings of sand etc. can be applied immediately after the extrusion; the system is particularly adapted for the production of perforated bricks.

In the past, bricks were left in covered rows – hacks – to dry in the open air; a procedure now discontinued, for obvious reasons, in wetter climates. Nowadays, the formed bricks are passed through heated chambers to remove a percentage of the moisture; during this stage they acquire firmness (green strength) so that subsequently they can be handled more readily.

The simplest method of **firing** bricks is a traditional one of long standing – the **clamp** method. In this the bricks are stacked together, allowing some air spaces between, while fine coal is sprinkled at intervals in layers. Formerly simply stacked in the open air, the modern clamp usually has a cover over the whole structure. The fuel is fired at one end, and the fire allowed to burn through.

Somewhat more complex is the **chambered kiln** – this consists of a series of brick-built chambers, generally arranged along the two sides of a rectangular structure. The chambers are connected by a series of flues; fuel is fed in from the top, and the fire allowed to move round the chambers in succession. The bricks are set in the centre of each chamber.

A **tunnel kiln** is a long, refractory-lined tunnel, through which carriages carrying packages of bricks slowly pass. In the centre of the tunnel is the firing area, into which fuel is fed from the top.

Whatever the type of firing, there is considerable variation in the setting of the bricks, and different patterns are used. A close setting, with little spaces, causes a slower penetration of heat to the centre of the package, and ensures that the centrally placed bricks are fired in a reducing rather than an oxidizing atmosphere. There is also scope to control this by regulating the overall supply of air to the kiln. It will be evident that each method of burning is devised to provide a period of pre-heating (the soak), followed by the intense burn, followed by a gradual cooling. The variation of all these factors allows the production of a wide variety of brick types. As will be seen in the sequel, the suitability of the firing process, and the effects that it produces, are largely determined by the original mineralogy of the clay; and thus study of the raw material is the first prerequisite in understanding the process of brick production.

5.3 THE CONSTITUENTS OF BRICKCLAY

With the exception of some clays of a residual origin, most brickclays are sedimentary deposits. They thus contain a variety of minerals, of differing origins, and in different proportions. Each of these minerals reacts differently to the various stages of the brick-making process; many of them of course interact in complex ways. Much remains to be understood; but in the sequel an attempt is made to assess the role of each mineral, at each stage in the manufacturing process.

Probably the most important constituent of a brickclay is, surprisingly, **quartz**. Quartz occurs as sand- or silt-sized particles in most brickclays; it can form as much as 90% of the total. Because it is not hydrophilic, its presence, in the sand grade, assists in separating the green brick from its mould. By creating a somewhat open texture to the brick, it assists the drying process, and is indeed sometimes added to a clay as a 'grog' (the industry's term for an inert additive).

Because it is relatively inert to all parts of the brickmaking process, the percentage of free quartz in a brickclay is an important parameter; since the quartz is generally of a larger grain size than the other constituents, particle-size distribution may be used as a crude measure of quartz content. This explains the frequently observed relationship of size distribution to

1. quantity of water required to make the brick (e.g. Figure 5.2)
2. shrinkage during drying (e.g. Figure 5.3), and
3. shrinkage during firing (e.g. Figure 5.4).

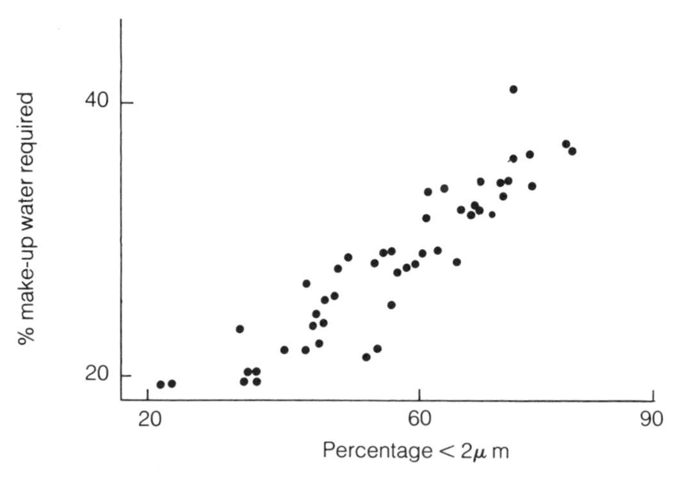

Figure 5.2 Relationship of the make-up water requirement of the clay to the content of fraction <2 μm. Wealden mudstones in the Hils (Lower Saxony, Germany). After Stein *et al.* (1980).

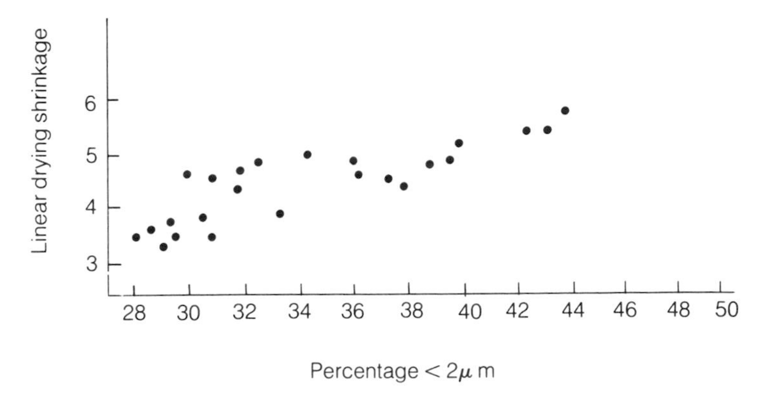

Figure 5.3 Relationship of the linear drying shrinkage of the clay to the content <2 μm. Jurassic mudstones from the area north of Osnabrück (Lower Saxony, Germany). After Stein *et al.* (1980).

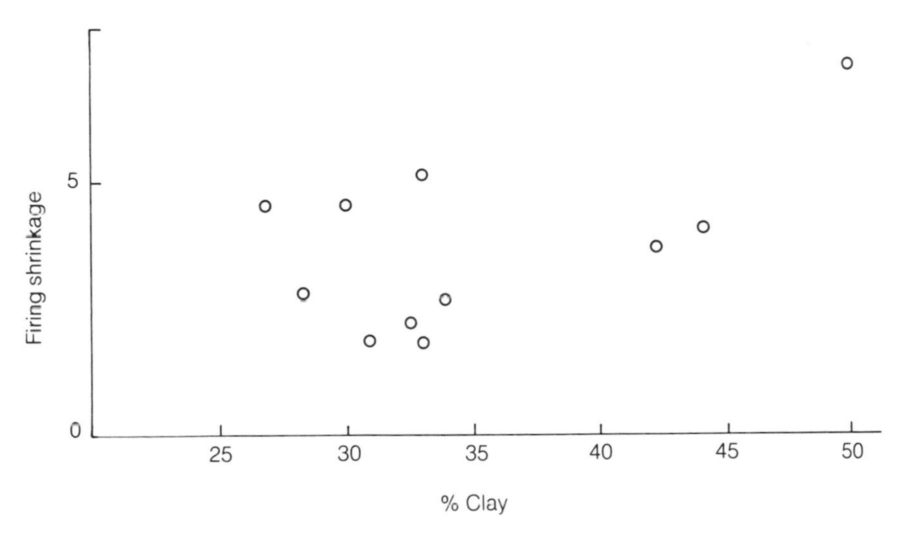

Figure 5.4 Relationship of the firing shrinkage of the clay to clay fraction.

During the firing process quartz undergoes two transformations – from a low-temperature (α) to a high-temperature (β) form at 574°, and from β-quartz to cristobalite at above 870°. The α–β transition involves a volume change of 0.8%, so that during the cooling of the brick microcracking occurs when this temperature is reached, thus lowering its strength. The β-quartz to cristobalite transition takes place slowly, so that unless the brick is held at high temperatures for a long period, only the smaller particles, or the edges of larger particles, are affected; they form then part of the glass phase (see below). Quartz is strong and resistant to weathering, and its presence in the fired product is largely responsible for the strength and durability of the brick.

The **clay minerals** are essential to brick and tile manufacture because they provide

1. the plasticity necessary to mould the brick
2. the ability to hold the form of the brick during the drying process, i.e. the green strength, and
3. the fusibility to form a glass, i.e. vitrification at relatively low firing temperatures.

Clays are sheet silicates, in which sheets of SiO_4 tetrahedra are combined with one or two layers of $Al(O/OH)_6$ or $Mg(O/OH)_6$ octahedra (Figure 5.5). Substitution by Al and other metallic elements is common, and interlayering of different clay minerals can also occur. In addition to this complex chemistry, identification is not simple, being dependent largely upon X-ray diffraction and scanning electron microscopy.

However, clays used in brickmaking are usually mixtures of no more than four clay minerals; these are **kaolinite**, **illite**, **smectite** and **chlorite**, as follows.

Kaolinite has a structure in which a single silica tetrahedral layer alternates with a single alumina octahedral layer, giving it a composition of $Si_4Al_4O_{10}(OH)_8$, and the highest alumina content (39.5%) of any clay mineral. In most sedimentary clays, kaolinite occurs in disordered form, in which there is a random element in the structure, accompanied by ionic substitution by Mg, Fe^{2+} and Ca^{2+}.

Illite has a structure in which an octahedral alumina layer is sandwiched between two tetrahedral silica layers; some of the Si^{4+} ions are replaced by Al^{3+} ions, and the sheets are bonded together by potassium ions.

Smectite is a name given to a group of minerals in which the oxygens are partially replaced by hydroxyls, and the interlayer cations are sodium and calcium. These minerals are characterized by their ability to take in additional layers of water between each unit sheet and the next.

Chlorite consists of illite-like sheets alternating with octahedral layers of composition $(Mg, Al)_6\,(OH)_{12}$. Sometimes this latter layer is discontinuous, which gives the mineral swelling properties – this is known as swelling chlorite.

Other minerals which sometimes occur in sufficient proportions

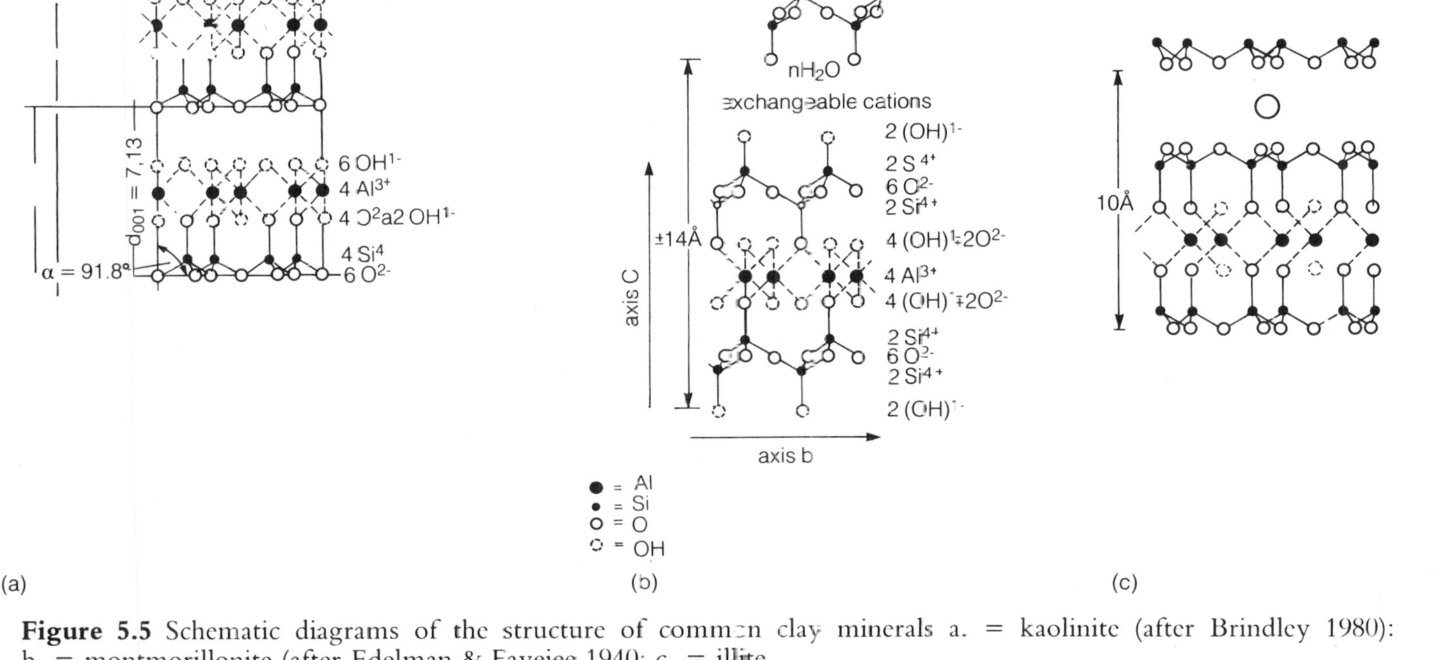

Figure 5.5 Schematic diagrams of the structure of common clay minerals a. = kaolinite (after Brindley 1980): b. = montmorillonite (after Edelman & Favejee 1940: c. = illite.

to affect the brickmaking properties are **micas**, **vermiculite** and **sepiolite**. Each of these minerals behaves differently at the various stages in the production of the brick. In the forming stage, illite and disordered kaolinites are plastic, whereas ordered kaolinites and chlorites are not. Smectites are thixotropic; they require a high addition of water to render them plastic, and this is lost in the drying process. They thus have a higher drying shrinkage than the other clays (up to 23% linear, compared with 10% maximum for kaolinite and 11% for illite). Smectites also tend to reabsorb water at the green stage.

All clay minerals, when heated to temperatures in excess of 1200°, are capable of being recrystallized to form the minerals **mullite** ($xAl_2O_3 \cdot ySiO_2$), **corundum** (Al_2O_3), and, where Mg is present, **olivine**; while cristobalite results both from changes in the mullite composition, and from the incorporation of free quartz. However, the temperatures and times used in industrial brick production are inadequate for these changes to go to completion, and fusion usually ceases at a stage of glass formation in which some of these minerals are beginning to form. Nevertheless, the development of the felted crystals of mullite is believed to be important in the production of strength in the fired brick.

Ordered kaolinite has a high fusion temperature (in excess of 1200°), and passes through no glass phase, but it can be shown that mullite does form if a temperature of *c.*950° can be maintained for more than a day. Disordered kaolinites have lower fusion temperatures, while illites readily form a glassy phase at around 1050°. The behaviour under firing of smectites and chlorites varies with their chemical composition. These reactions frequently involve some movement of iron, thus affecting the colour of the fired brick; the kaolinite/mullite tranformation can take in some iron, thus bleaching the brick, while iron is expelled from illites and chlorites at an early stage, causing reddening: It follows that the firing schedule of each brick needs to be carefully adjusted to the clay-mineral content of the raw material, if a brick of the required strength, durability and appearance is to be produced in the most economical way.

Iron minerals provide most of the colour of the fired brick. In the raw clay, the commonest iron minerals are **haematite** (Fe_2O_3); **goëthite** ($\alpha FeO \cdot OH$), **limonite** ($\simeq 2Fe_2O_3 \cdot 3H_2O$); **magnetite** ($Fe_3O_4$); **pyrite** ($FeS_2$) and **siderite** ($FeCO_3$). Under oxidizing conditions in the kiln, all these convert to haematite, which is of paramount importance in producing brick colour. Haematite, on heating above 1000°C, shows increasing lattice disorder, and becomes increasingly darker red in colour; thus in general a higher firing temperature produces a redder brick. Reducing conditions can be created in the kiln by the bricks being set close together, so that air passage between the bricks is restricted; or by controlling the fuel supply so that all the oxygen is burned. In a reducing atmosphere, the iron combines with the silicates in the clay to form ferrous silicates, which, unlike haematite, become liquid at kiln temperatures, thus forming a dark blue skin on the surface on cooling. This is the basis of the manufacture of the Staffordshire Blue brick, made from the iron-rich Etruria marl of the West Midlands of England. These conditions are sometimes combined; by a process known as flashing, fuel is increased for a short period at the end of the firing cycle, thus elevating the temperature and reducing the oxygen – by this means a varied suite of colour can be produced on the brick surface.

Reducing conditions in the interior of the individual brick can be created when the rise in temperature of the kiln is such that a vitrified skin forms rapidly on the outside of the brick, thus preventing the escape of gases. This produces a black core to the brick. In some cases, the black core extends to very near the brick surface, and is deliberately created to give colour effects; but its presence tends to weaken the brick.

Thus by careful control of the kiln temperature schedule, and the atmosphere within it, the brickmaker can produce a multiplicity of colour variations. However, these changes are accompanied by numerous other mineralogical changes, which can exert subtle influences on the colour, so that precise control of colour is in fact extremely difficult; the complexity is well illustrated by Figure 5.6.

Some brick clays produce white, yellow or buff coloured bricks on firing, and the mineralogy of this process is complex. In kaolinitic clays, which are deficient in alkalis (Figure 5.7), the kaolinite is first transformed, at around 1000°C, into amorphous metakaolinite, and then into crystalline mullite (Al_2SiO_5); both these minerals are able to take iron into their lattice in substitution for aluminium. This can be demonstrated in the crystalline mullite by measuring the lattice dimensions (Figure 5.8), which increase with increasing iron substitution. It will be seen that up to 11% of haematite can be taken in, so that up to this level the bricks are pale in colour. Iron oxide in excess of 11% crystallizes as haematite, and the bricks are reddened. The presence of calcium is also important. Low percentages of calcite (2–10%) result in the formation of anorthite, which is unable to accept iron substitution, so that the haematite is freed to give a red colour. At higher levels of calcite (around 20%), some of the iron is taken into a pyroxene, which is brown in colour, and such bricks

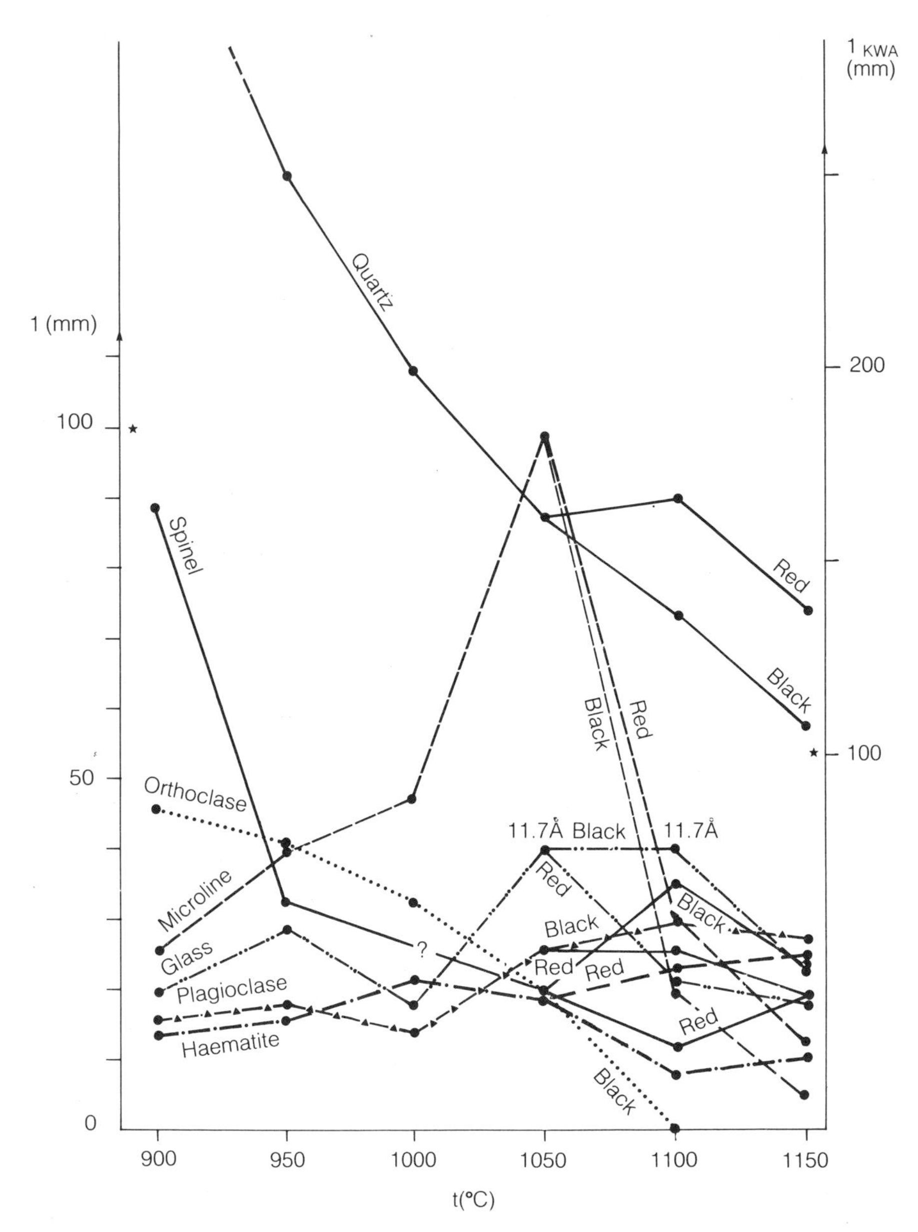

Figure 5.6 Mineralogical changes in brickclay during firing. Sample of Boom clay (Miocene, Belgium). Vertical peak is peak height of X-ray trace. Above 1050°C separate analyses for red (oxidized) and black (reduced) parts of the sample; note the changes in haematite/spinel (magnetite) ratio above this temperature. After Decleer (1983).

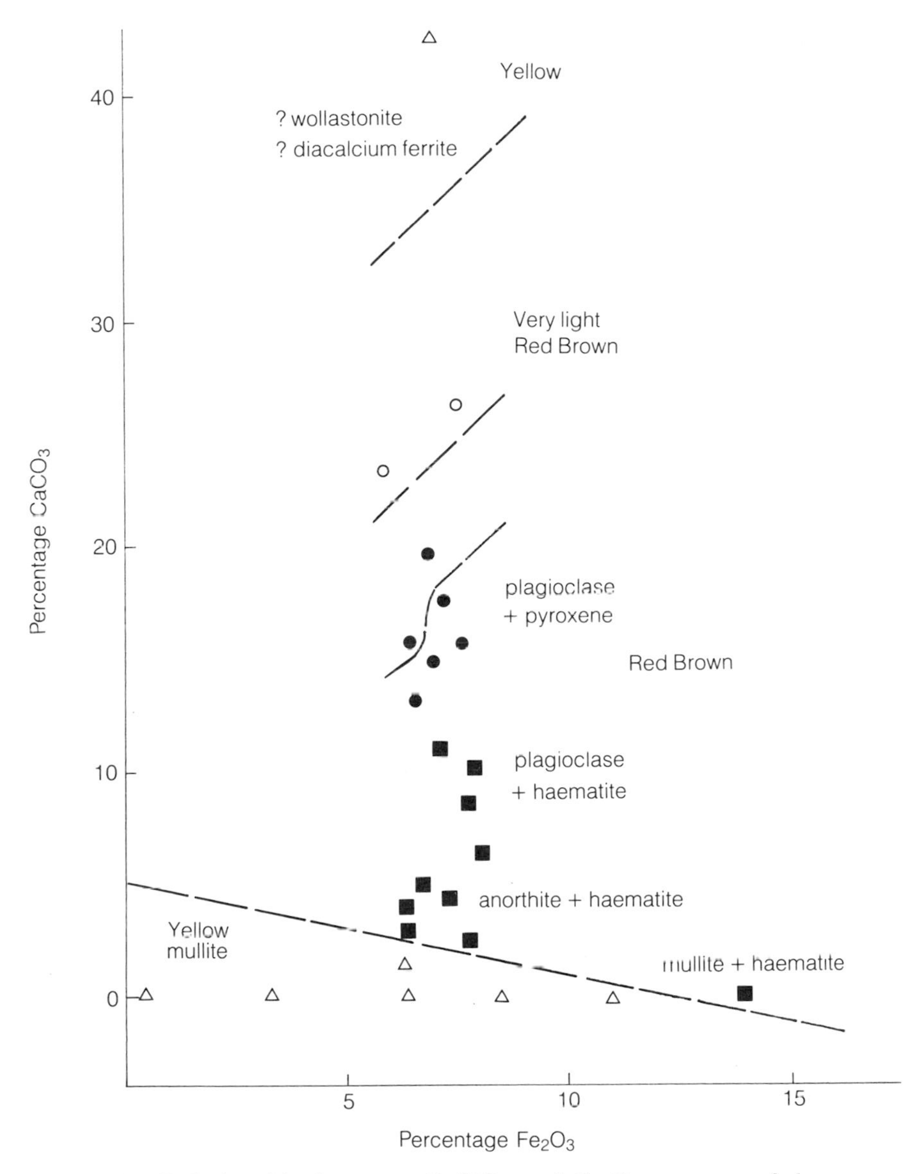

Figure 5.7 Relationship between $CaCO_3$ and Fe_2O_3 content of the raw material and the fired colour in kaolinitic clays. Data from Kreimayer and Eckhardt (1987) and Stein *et al.* (1981).

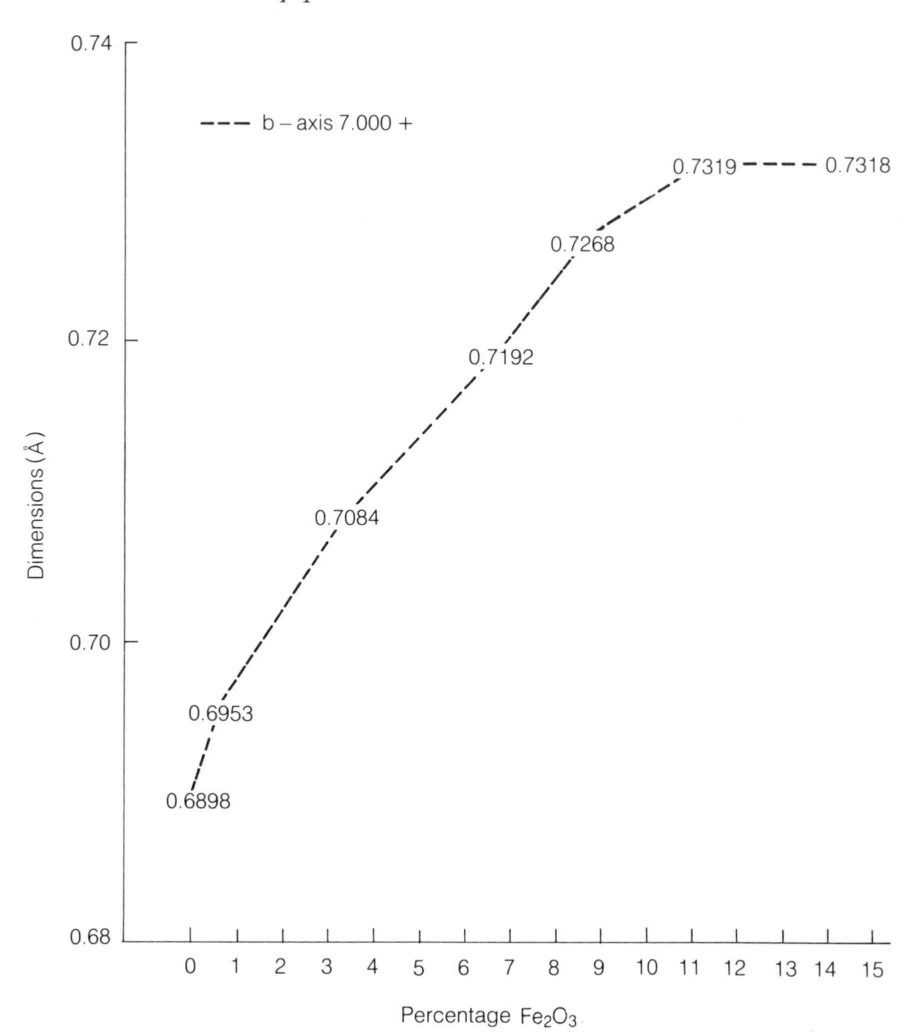

Figure 5.8 Relationship of *b*-axis lattice dimensions of mullite to iron content of a brickmaking clay.

are described as 'light red-brown'. Higher concentrations of calcite cause a progressive lightening of colour, so that at 35–40% $CaCO_3$ the bricks are again yellow in colour. This appears to be due to the production of a dicalcium ferrite ($2CaO \cdot Fe_2O_3$): **wollastonite** ($CaSiO_3$) is also known to be formed at these high calcium levels.

These reactions only go to their full extent when the constituent particles of the brick clay are finely divided. When they are coarse, reactions are incomplete, and the mineral **gehlenite** is produced; this can incorporate iron but gives a dingy colour to the brick.

The presence of alkali elements (which will chiefly be in the form of the clay-mineral illite) has an important effect; in the presence of only one or two per cent alkali, anorthite and other plagioclase feldspars form instead of mullite, so that free haematitie occurs even when its concentration is low (<6%) and the bricks are reddened. The effect of the calcium presence is also enhanced, so that illitic bricks with only 15–17% of calcium carbonate and up to 6% iron oxide fire to a yellow colour.

The iron sulphide, pyrite, has a separate significance. Between 480°C and 588°C it evolves gases, either hydrogen sulphide or sulphur trioxide (depending on the amount of water and oxygen present); if at this stage the outside of the brick has become vitrified, a reducing environment is created, which forms the black core. In extreme cases this will cause the whole brick to expand, so that it inflates, and even explodes – a phenomenon known as bloating. Even if this stage is not reached, the spongy texture resulting from this process often reduces the strength of the brick. The presence of pyrite in a brickclay can therefore only be tolerated if the proportion is low. It is also very susceptible to weathering, so that it is not present in surface, weathered clay; and a short period of exposure, as, for instance, in a stock pile, will effectively oxidize it. Unfortunately, the products of such oxidation are also deleterious (see below).

Iron minerals can also act as a flux – i.e. as a cause of lowering the vitrification temperature. In the Carboniferous Etruria marl, for instance, in the absence of any other fluxing agent, their role in this respect is critical. Alternatively, where other fluxes are present, the maximum permissible iron content is quite low, because exceeding this causes the bricks to melt and deform in the kiln (squabbing).

Calcium minerals are usually present in clays in the form of **calcite** ($CaCO_3$); this occurs as fossil shell fragments, as concretions of various sizes, as thin seams or lenses of limestone, or finely disseminated throughout the clay.

On heating to 900°C, calcite evolves carbon dioxide, which contributes to the reducing atmosphere within the brick, and so to the formation of a black core. If calcite is present in quantity, the presence of large quantities of carbon dioxide in the kiln can have a damping effect, resulting in the bricks being underfired, and failing to develop their proper colour and strength.

In fact, the presence of calcite in the clay can have important effects upon colour development, and is sometimes added to normally red-firing clays to produce a yellow or buff colour (as for instance, in the Kent stock brick – see below). The mechanism of this is unclear, but probably involves the incorporation of some of the iron into complex carbonates such as **ankerite** ($(Ca \cdot Mg \cdot Fe)CO_3$), which do not have the strong colours of haematite.

If the calcite resists grinding, and therefore survives into the made brick in fragments larger than a millimetre diameter, the evolving gas creates a small bubble in the vitrified clay, and the calcite is converted to soft calcium oxide; if this occurs near the surface it can produce unsightly pits.

In pyritic clays, the sulphuric acid produced by the decomposition of the pyrite will react with any calcite present to produce **gypsum** ($CaSO_4 \cdot 2H_2O$). This can take place either during the weathering, natural or artificial, of the clay; or within the brick in the kiln. During firing it is converted to a hemihydrate form; but once the brick is exposed again to the weather this reverts again to gypsum. Evaporation of interstitial water from the brick brings this to the surface as a white efflorescence: at best this is no more than an unsightly scum which is soon removed by rainwater; but if much is present the crystallization effects within the brick can result in its total disintegration. Reactions between these soluble salts and liquids from the mortar can also be very harmful.

Carbon occurs in three forms in brickclays: as **coal** or **lignite** or as **hydrocarbon**. In the first form it is found chiefly as fragments of plant debris, rootlets, or fossil wood, although sometimes thin seams of coal are deliberately included in the mix. While reducing the fuel cost of firing the bricks, this often has the effect of bleaching the colour. In its hydrocarbon form, as found in oil shales, it can be most valuable, reducing energy costs significantly; but there are upper limits beyond which it is not possible to control the burning process (see below).

Most brickmakers make use of **additives** to change the properties of the fired brick. In general, the low selling price of the product prevents any expensive or bulky additions to the raw clay; for this reason most additives are common industrial minerals or waste products of other industries. They serve only therefore to modify the original clay, whose characteristics are thus still largely determined by the quarried material.

Grog is a name given to any more or less inert material, added to open out the texture of the brick, and thus to reduce risks of bloating and black coring. Sand is the most common, but crushed rock such as mica-schist, volcanic ash, and porphyry, as well as waste products like incinerated domestic refuse, and recycled burnt brick are used – dependent very much on local availability.

Colorants may be incorporated in the body of the brick, or in the sand used for demoulding or facing. Those used are the cheaper industrial stains, such as the ochres, and manganese dioxide.

It is common practice to add some fuel to the brick to produce even firing; coke is the preferred material, since it also serves as a grog; but powdered coal, waste oil, etc. have all been tried. Sometimes a non-mineral addition is used – for example sawdust, or the waste fibre from sugar-cane processing (bagasse).

Where sulphate scum is a problem, the addition of barium compounds is effective as a control; this precipitates the sulphate as the very insoluble barium salt, which therefore does not move, or recrystallize, within the brick. Lime pitting is sometimes reduced by the addition of common salt – although how this works is still unclear.

5.4 OCCURRENCE AND DISTRIBUTION OF BRICKMAKING CLAYS

Clay minerals are formed by the weathering of aluminosilicate minerals; since these are present in almost every rock (except pure sandstones and limestones) it follows that clays are very abundant. Thus it is possible, almost everywhere in the world, to make a brick from weathered material scraped from the surface.

The kind of clay mineral produced by weathering is closely dependent upon the climatic conditions, and especially on rainfall (Figure 5.9). It will be seen that dry conditions favour the development of smectite; illite forms under somewhat wetter conditions, while very wet conditions produce kaolinite and vermiculite. In areas of highest rainfall, silica may be removed from the molecule, producing gibbsite ($Al(OH)_3$). The last reaction is enhanced by high temperatures, and so the very large areas of the globe which have a hot, wet climate have a thick weathering crust in which gibbsite and kaolinite predominate – a laterite.

Sedimentary rocks of course are formed by the removal and redeposition of the weathered surface material; and during the process of sedimentary transport clay minerals appear to be remarkably stable. Thus the clay sediments of the Atlantic adjacent to tropical Africa and South America are dominated by kaolinite. At this stage, clay minerals from other sources than weathering may be added; thus the large areas of smectite in modern Pacific Ocean clays are attributed to a volcanic origin; while the abundant chlorite off the Scandinavian coasts is thought to have been formed by mechanical weathering of pre-existing chloritic shales.

Some changes take place during diagenesis; kaolinite appears persistent in lake and river environments, but in sea-water may be transformed to illite; smectite may also be changed to illite at this

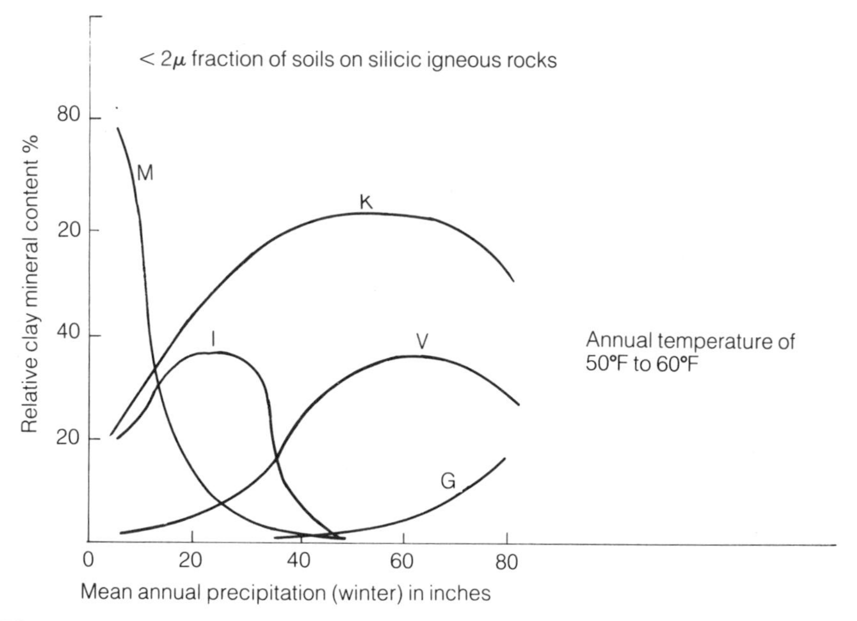

Figure 5.9 Variation in clay mineralogy with climate in soils developed on silicic igneous rocks in California (after Barshad (1966)).

stage. Further transformations to illite and chlorite take place with shallow burial and temperature rise.

It follows that the precise clay mineralogy of a brickmaking clay is closely determined by its geological history; Figure 5.10 shows how these clay minerals vary with the age of the rock. Notable features are the increase of chlorite and decrease of kaolinite with age; and the almost total change of clay mineralogy in the arid conditions of the Permo–Triassic.

Other, non-clay minerals are clearly related to the mode of origin of the raw material. Thus in laterites, and in laterite-derived sediments, the iron is mostly in the form of haematite: in non-marine sediments it is as siderite; and in marine sediments, particularly those formed in deep water, it occurs as pyrite.

5.5 MAJOR BRICKMAKING CLAYS – BRITAIN AND WESTERN EUROPE

The use of surface, weathered clays as a raw material for brickmaking continued well into this century in Europe and North America, and is still important in many developing countries. In the developed countries, however, the small local brickworks have

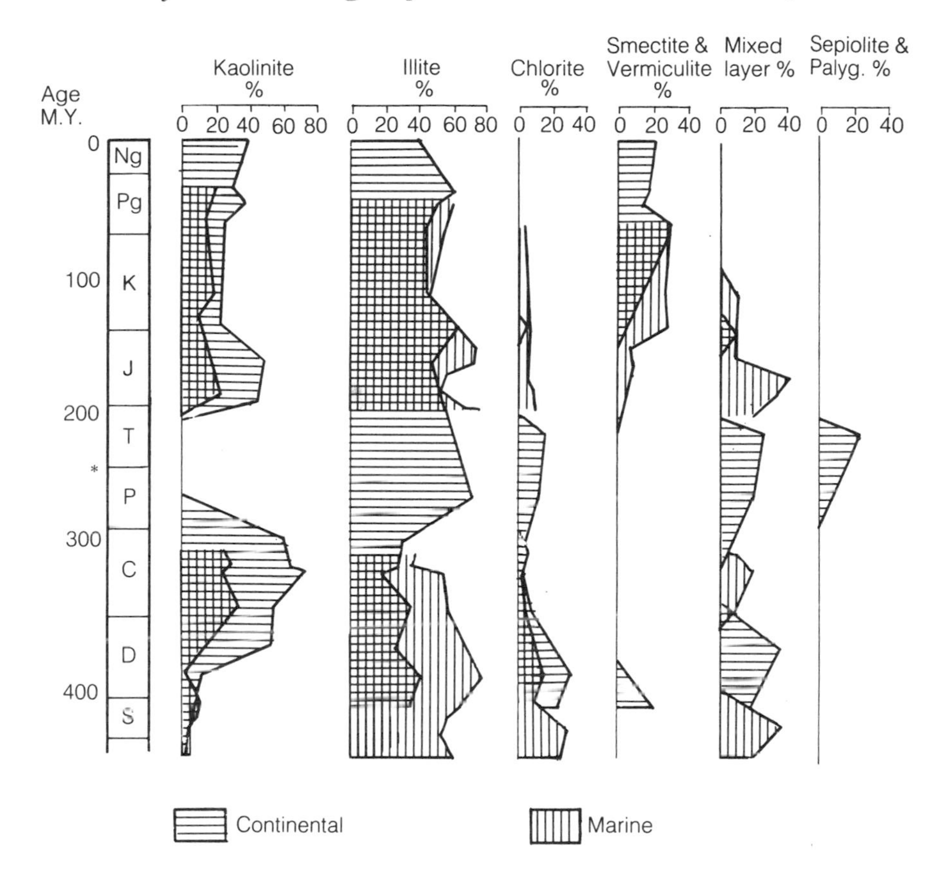

Figure 5.10 Variation in abundance of clay-mineral species with time (after Ridgway (1982)); data mainly from Perrin (1971).

gradually been reduced to a few very large units; these must perforce look to deep, unweathered material for their supply; and need to find geological formations which are thick, consistent both laterally and vertically, and which are to a large extent free from manufacturing problems. To these constraints must be added the demands of the market, which sometimes favours excessively a brick whose raw material is scarce, and will not buy a brick made from an abundant source. All this has led to a steady reduction in the number of geological formations which are still used; in Britain, for instance, it now amounts to a dozen or so only (Figure 5.11).

Very few pre-Devonian rocks are suitable for modern brick-making; they have mostly been involved in at least one phase of deep

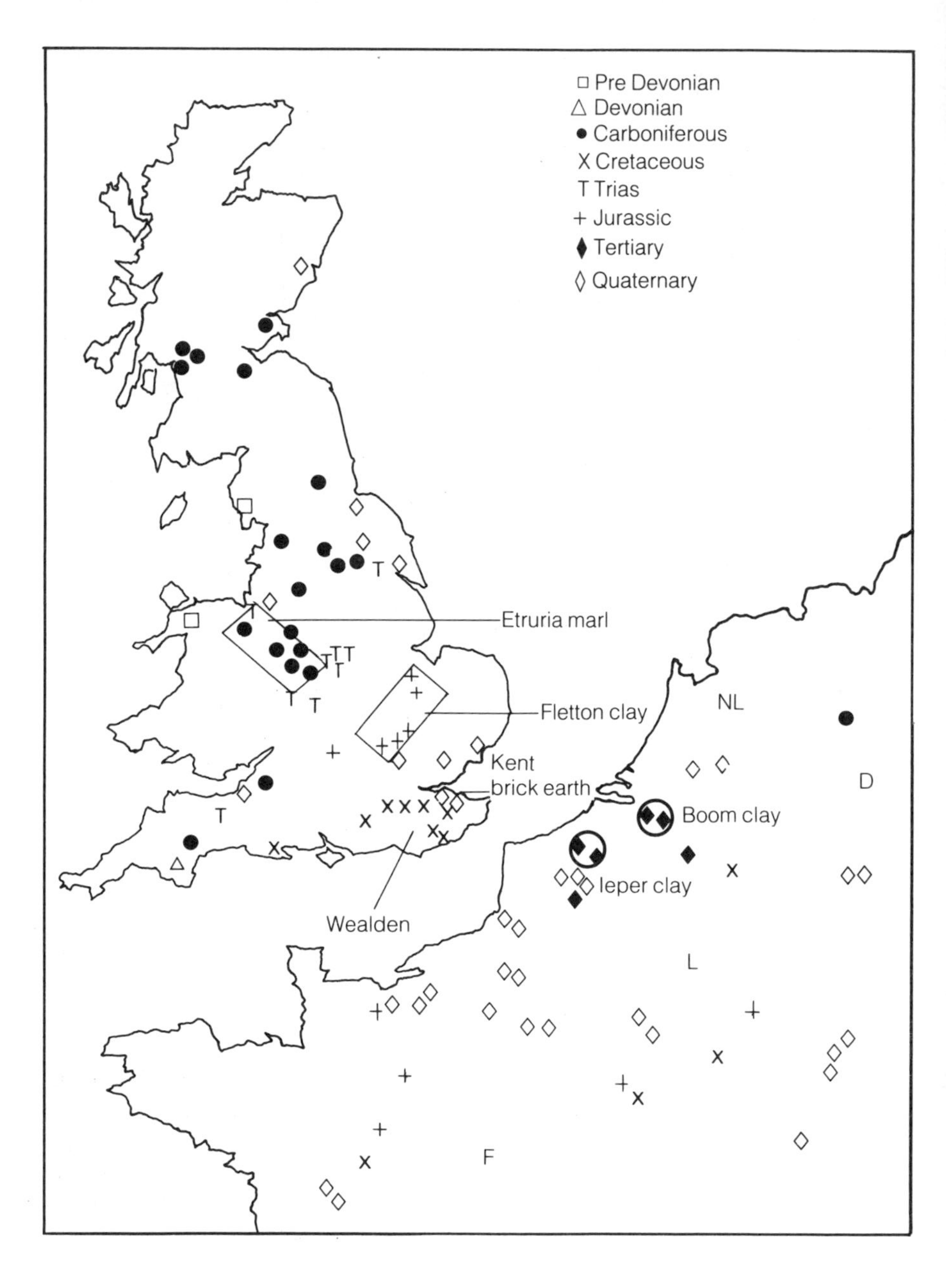

Figure 5.11 Distribution of brickmaking clays in north-west Europe.

burial and compression, and so are usually highly indurated, which adds very substantially to the cost of quarrying and grinding. The clay mineral is described as illite (but often of well-ordered type approaching mica) and chlorite, and imparts little plasticity to the clay; but both these minerals form glassy phases from 950°C onwards, so that firing temperatures can be quite low.

In Europe, the Devonian occurs in two facies; as marine slates and sandstones in the south and the non-marine Old Red Sandstone in the north. The former were involved in the Variscan orogeny, and from the brickmaking viewpoint, suffer from the same disadvantages as the Lower Palaeozoic rocks; they are, however, used with **ball clay** (see below p. 167) as a plasticizer in south Devon. The Old Red Sandstone, despite its name, consists very largely of red marl; and not having been involved in any major tectonism, has retained its plasticity and has proved a satisfactory brickmaking clay.

The coal-measure facies of the Carboniferous is of deltaic origin; lying offshore from a continental mass subjected to tropical weathering, it contains a high proportion of kaolinite; and this has been partly converted under conditions of shallow burial to illite and chlorite. This favourable combination of clay minerals has ensured that the coal measures are a major source of brickclay throughout north-western Europe. Another reason for this is economic; from the industrial revolution onwards it was realized that brickclay extraction could be carried on in conjunction with coal mining, particularly if the latter was by opencast methods. Thus many brickworks were based on raw materials not selected for their brickmaking properties, and many such brickworks existed which produced inferior bricks. The rapid variations of lithology, both vertically and laterally, in the coal measures have made it necessary, if good bricks are to be produced, for selective extraction and careful blending to be carried out. To control this selective extraction, a proper understanding of coal-measure sedimentology is essential.

In recent years, there has been a concentration of brickmaking on to a few of the most suitable parts of the coal-measure sequence. Thus in northern England, the Accrington mudstones – a 30 m thick sequence of mudstones in the middle coal measures – has been intensively exploited. It is very high in illite, and produces an evenly coloured red brick of high compressive strength and great durability.

Within the coal-measure sequence, generally immediately below coal seams, there occur **fireclays**; these are fossil weathering horizons leached by acid percolation; they are predominantly kaolinite, but associated with other high-alumina minerals; and they are very low in iron. They are highly refractory, and at the firing temperatures used for brickmaking cannot easily be used on their own; but they can be mixed with other clays to produce a buff-coloured brick.

During the late Carboniferous, the surface of the continental masses became intensively lateritized, and the coal-measure delta received floods of red sediment. A sequence of red clays and sandstones formed in this way, and extensively exploited by the brick industry in north-central England, is the Etruria marl. In these clays kaolinite is always in excess of illite, and free quartz may constitute as much as 45%; there is a virtual absence of pyrite, gypsum and calcite. Haematite may be up to 9%, and in the firing process appears to act as a flux. Not only does this clay produce a strong, evenly coloured red brick, but also under reducing conditions it gives a heavy, dense, blue-black brick, which as the Staffordshire Blue found great favour with the engineers of the railway age.

Desertification during the Permian and the Triassic periods resulted in clay formation on the land masses in which little kaolinite was produced. Although thick clay formations exist across north-western Europe (e.g. the Mercian mudstones of Britain, Keuper marl of Germany, etc.) the number of large brickworks based on this is relatively small probably because the clays present considerable processing problems. They contain a high proportion of fine-grained quartz, originating as wind-blown dust, so that quartz inversion and cristobalite formation become significant factors. Carbonate contents are generally high (up to 35%), often as fine-grained dolomite; this lowers the firing temperature; but rapid local variation in carbonate content causes very wide colour variations, from buff to dark red. The clay mineral content is dominated by smectite and illite, with little kaolinite, so that the vitrification range is short, and therefore difficult to control. Locally the clay minerals include magnesian varieties such as sepiolite, which further modifies the brickmaking process. In general, bricks made from these formations are rather less dense, more porous, and less durable than those made from other clays; and most brickmakers need to blend clays, either from within the formation, or from other geological horizons.

The beginning of the Jurassic saw a marine transgression over much of northern Europe, and a return to wetter conditions over the land. However, the land masses had been much worn down since their latest elevation in the Carboniferous, so that the seas were mostly free from sediment; so much of the Jurassic and the Cretaceous rock consists of limestone. There are, however, some thick clay formations which are important sources of brickmaking clay.

Clay horizons occur at various levels, particularly in the lower parts, of the Liassic in Britain, northern France, Luxembourg and Germany. The clays are dominantly illitic, and can produce strong, dense bricks of good colour and durability. Their main disadvantage is in the presence of calcite; this occurs as discrete limestone bands (which can be removed by selective extraction), and as very abundant fossil shells. The clay also may contain pyrite, and carbonaceous fragments; so that it is far from ideal as a brickmaking material.

By contrast, the Upper Jurassic sequence contains a uniquely favourable brickmaking clay, the basal 70 m of the Oxford clay of Britain (the Fletton knotts) which forms the basis of an industry which accounts for more than half the volume of output of all British brickmakers. Illite is the dominant clay mineral, with a smaller percentage of kaolinite; chlorites, smectite and vermiculite are also present. Carbonate contents are low, being produced mainly by the shell fragments of fossil ammonites and bivalves, which in the relatively deep water in which these clays were deposited, all have thin shells and are relatively sparse. Pyrite is present, but often only as large concretions, or concentrated in specific layers, notably towards the base. But the real advantage derives from the presence of very finely divided hydrocarbon. This gives an enhanced plasticity to the ground clay; the clay retains some 18% water and therefore can be pressed easily by a semi-dry process. The presence of the contained fuel reduces the cost of firing by as much as 75%; so that bricks made from Fletton clay can be sold at much lower prices. Set against this is a low density and high water absorption, and in their unmodified state, a not very attractive body colour – which can be successfully masked by surface treatment for use as a facing brick. Curiously, the clays with high calorific value do not appear to extend outside Britain; and works at this geological horizon in Normandy and Germany do not have the advantage of highly calorific clay.

Another clay formation higher in the sequence, the Kimmeridge clay, is somewhat similar to the Fletton clay, and has been used in the past; but carbonate contents are higher, and in many cases the bitumen content appears to be too high.

In the Cretaceous, the Wealden clays of south-eastern England are an important raw material. They were deposited in a shallow lake, or arm of the sea, adjacent to land masses subjected to a warm wet climate; kaolinite and illite are in roughly equal proportions, and quantities of smectite are low. Perhaps most important, there are wide variations in the content of various ancillary minerals, so that the formation produces a wide variety of highly individual bricks – an important asset in a region adjacent to a very sophisticated

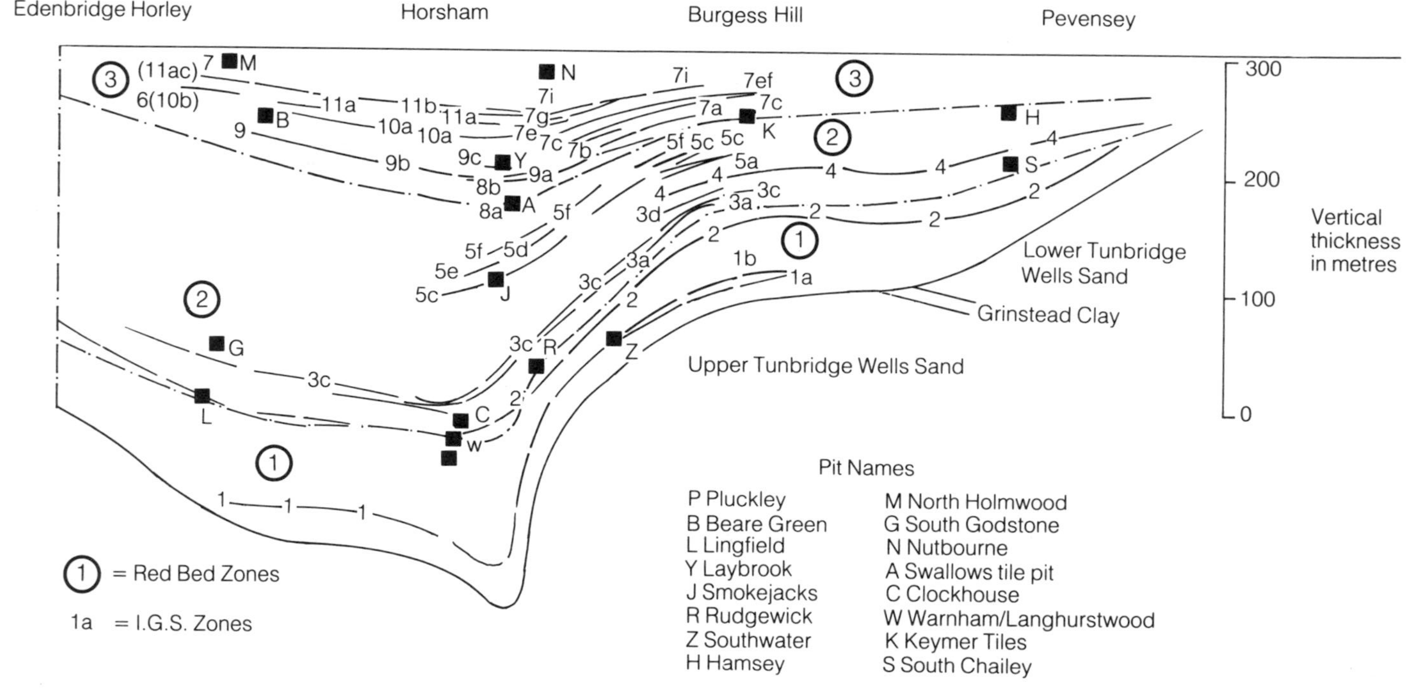

Figure 5.12 Stratigraphic section across Weald, showing location of brick-works in relation to stratigraphy of Weald clay.

market. Thus there are many brickworks situated at different stratigraphical levels within the sequence (Figure 5.12). The Wealden of Belgium and Germany is a much more sandy formation.

In strong contrast to the Weald clay, the Upper Cretaceous Gault presents many problems to the brickmaker. A marine clay, of deep-water origin, it has a high clay fraction, much of which is smectite, and which creates difficulties in forming and drying; it contains abundant pyrite, with consequent problems of bloating and efflorescence, and the high carbonate content tends to produce a yellow brick. While formerly much used in its weathered state, the only surviving users are based on some limited horizons in the sequence where deleterious minerals are at a minimum.

The complex of marine and non-marine clastic sediments which make up the lower Tertiary of south-east England and the Paris basin were exploited at many stratigraphic horizons by a host of small brickworks in both countries; but rapid local lithological variations and dwindling reserves have resulted in the closure of most of these. In Belgium, however, there are thick, laterally extensive bodies of marine clay which form the basis of a large-scale industry. It is characteristic of these clays that smectite content is high – this is attributed to soil formation under semi-arid conditions on the adjacent Scandinavian land mass; there is a systematic change to more illitic–kaolinitic clays through the Tertiary. In West Flanders, the Ieper (Ypres) clay is remarkable in that 90% of the clay mineral is smectite; but the quantity is consistent, and careful drying and firing schedules enable good quality bricks to be produced. Further north, in the Antwerp region, the late Oligocene Boom clay is a deposit unique to Belgium. This is a smectite–illite clay, with only subordinate kaolinite; a 50+ m thickness of rhythmically deposited clays and silty clays can be effectively mixed to give a consistent product. The presence of calcite and pyrite is apt to cause bloating, however.

The most extensive clay deposits of the Quaternary – the boulder clays – are almost useless for brickmaking, since their stone content cannot be separated, and even if they can be processed without damage to the plant, their very mixed nature makes a very inconsistent product. Most of the Quaternary deposits now used are thus either lake deposits or wind-blown loess.

Filled lake-basins, of both interglacial and postglacial age, form the basis of brickworks in Britain, Belgium, the Netherlands and Denmark. The clays are generally silty, often rhythmically banded (and therefore readily mixed), and, because of the quiet-water conditions of deposition, often very consistent in composition. Clay mineralogy is often closely linked with the local source area – thus Belgian examples near the Tertiary smectitic clays are also smectitic, whereas in north of England examples, chlorite predominates. Such deposits are, however, inevitably limited in extent by the confines of the original lake basin.

In contrast, wind-blown loess deposits of late Pleistocene age occur in extensive sheets, and often reach hundreds of metres in thickness. In south-eastern England, in Kent and Essex, we are at the northern limits of one such sheet; and although the Kent brickearth deposits are at most 2 m in thickness, they have formed the basis of an industry producing the London or Kent stock, whose characteristic yellowish-grey colour is evident throughout London building. The loess is mainly finely divided silica, with only a small proportion of clay minerals, and in the past decayed domestic rubbish, cinders and recent Thames river mud have been added. The high silica content is believed to be the reason for this brick's durability in the smoke-laden atmosphere of nineteenth century London. The yellow colour is related to the presence of calcite, generally occurring in bands of small concretions – **race** – and their sporadic occurrence gives a great variability in the calcium carbonate content. Fortunately, the nearby availability of ground chalk allowed this to be added to maintain consistency of colour.

In France, the much thicker limon which extends across the Paris basin and down the major valleys to the south supports a much larger industry, with many large and technologically advanced units producing many millions of bricks. This material has proved particularly suitable for the manufacture of roofing tiles, and the presence of these deposits in quantity is probably one of the major reasons that the clay-tile industry of France has survived in competition with the concrete tile for roofing purposes.

5.6 OTHER AREAS

The brick industry of the world is so widespread, and so varied, that a comprehensive review is not possible here. Everywhere in the world, some local material is obtained to make a brick. For example, in Georgia, USA, a mixture of pure kaolinite, with a little feldspar, bentonite and fireclay – with no quartz at all – is made into a strong white brick. In Malaysia a brick is made from weathered granite, laterite and tin-mine tailings. It seems that as long as some small proportion of clay mineral is present, some kind of brick can be produced.

The brick industry of the United States is an interesting illustration of the importance of clay-mineral types. The east and north of the country has a situation very similar to that of Europe, with a wide variety of available sources, ranging from Palaeozoic shales, through glacial clays, to recent loess – most of these having a mixed mineral composition including illite, smectite and kaolinite. In these parts there is a thriving brick industry. In the west, however, extensive tectonism has converted most of the clays to chlorites and micas in the older rocks, and the predominance of vulcanism in recent rock formation has produced clays which are largely smectite. So today there are, for instance, no brickworks at all in Montana, and two only in each of the states of Oregon and Idaho.

On a world-wide basis, however, probably the greatest volume of bricks is made from the recent alluvium of the world's major rivers. For centuries, the peasant farmers of the Yangste Kiang, the Indus, the Ganges, the Nile, the Mekong and so on, have scraped the top few feet from the flood plains of their rivers, built and fired the bricks in clamps, and made bricks. Such bricks are usually of low clay content, so that they dry easily, and on firing to fairly low temperatures, produce a weak and porous brick. But for one-storey houses, and for regions where the climate is relatively non-aggressive, such bricks are perfectly adequate.

5.7 VITRIFIED CLAY PIPES

Pipes for underground sewers need to be protected from attack by groundwater from outside as well as from the liquids within. They need a high compressive strength since they are often below roads and need to carry the loads of deep burial beneath high buildings. Size and shape tolerances are very strict, and to avoid distortion of such large objects in the drier and the kiln a well-mixed and consistent clay supply is demanded. In Western Europe the kaolinite/illite Westerwald clays, often grogged by the addition of ground-fired clay (chamotte), or local sandy clay, are favoured, and sometimes transported long distances: in England and North America carefully blended ground coal-measure shales are used.

To create the high strength, kiln firing temperatures reach 1200°C, so that crystalline secondary mullite is formed. So important is this mineral, that it is proposed to include mullite percentage as a test in the new German DIN specification.

The traditional salt glaze, produced by liberally sprinkling the surface of the formed pipe with rock-salt before firing, has now been almost entirely replaced by a feldspar-based ceramic glaze (Figure 5.13).

Figure 5.13 Production of 200–800 mm vitrified clay pipes at Keramo N.V., Hasselt, Belgium. Pipes are extruded and set vertically, and glazed immediately after extrusion, by equipment developed and built by Keramo (photograph courtesy of Keramo N.V.).

5.8 FLOOR AND WALL TILES

Decorative tiles for indoor use are commonly made of a clay-based body to which a silicate glaze is applied. Traditionally, the base is fired first, the glaze and any pattern added in liquid form, and then the product is fired again – a process which may take several days to complete. Modern methods, using a single firing, can complete the process in 20 minutes. In these circumstances, firing shrinkage needs to be very low, and very uniform; for this reason, refractory clays with minimal fluxing agents, finely ground and extremely evenly mixed, usually form the body. A typical recipe would consist of 35% ball clay, 15% china clay, 10% limestone, and up to 15% ground tile waste. Silica forms 25% of this mix, some as finely ground sand but frequently including a percentage of cristobalite (manufactured by calcination of sand) – this to avoid the effects of the α/β-quartz inversion at 573°C.

The glaze is essentially silica, with nepheline syenite or feldspar added as a flux; this is coloured by metallic oxides and fused in a furnace. Dropping the melt directly into water produces a frit – a highly porous glass which can be readily remelted to apply to the tile surface. Certain metal oxides act as 'opacifiers; in the past lead and tin were used, but these have been largely replaced by zirconia (derived mainly from black sands).

In the last century opulent public and private buildings frequently used encaustic tiles for flooring, and there is now a revival of their manufacture. For these a 2-cm-thick ball clay/nepheline syenite body is impressed with a deep pattern, which is then infilled by liquid clay slip of various colours. After very prolonged, slow drying the surface of the tile is ground flat, and then fired in the normal way.

In Britain, clays for ceramic tiles come mainly from south-west England. The **primary kaolinites** or **china clays** derived directly from the *in-situ* alteration of the granites of Devon and Cornwall, are in general too costly, and too refractory, for the tile industry, but the **ball clays** which occur in sedimentary basins in Devon and Dorset (Figure 5.14) are used. These consist of kaolinite which has a lower crystallinity, together with other clay minerals such as illite, so have a lower firing temperature than china clay. They are low in quartz and almost free from iron minerals, so fire to a white colour. These ball clays owe their existence partly to the reworking of the nearby altered granites, and partly to the destruction of an extensive kaolinitic weathering mantle, which developed over England in the hot humid conditions of the early Tertiary, and which has been almost entirely removed by the deep erosion of late Tertiary and Pleistocene times. Similar deposits occur in the Westerwald area of Germany, where they form the basis of ceramic tile manufacture in that and neighbouring countries.

Clays used in Italy and Spain, however, are much more variable. Kaolinite does not usually make up more than 25%, and is often completely absent; so that illite is the dominant clay mineral, with

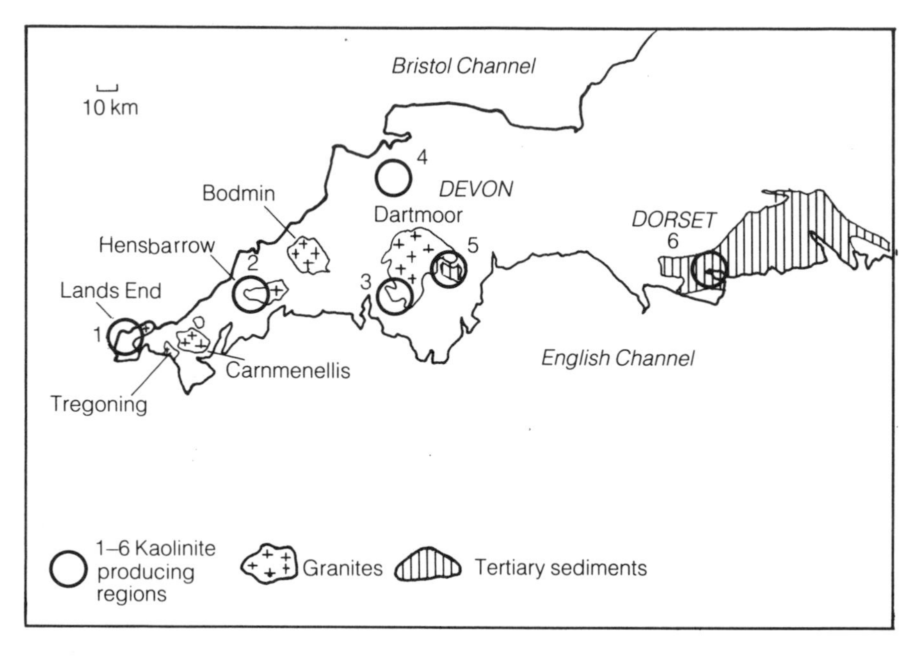

Figure 5.14 Distribution of primary and secondary kaolinite deposits in south-west England.

significant quantities of smectite and chlorite. Iron content may be as high as 6%, so that bodies are often quite dark red in colour. Quartz contents are also quite high (up to 40%); and there is often a high content of carbonate minerals, both as calcite and as dolomite, either of which can reach 18% in the body mix. Such tile bodies are inevitably weak, and there is an increasing tendency to add felspathic minerals, or kaolinitic clay imported from England or Germany, to add strength to the tile.

In North America extensive kaolinite deposits occur in South Carolina and Georgia. The crystalline rocks of the Piedmont region, during a long period of crustal stability from Triassic times onwards, developed a deep kaolinitic clay mantle, which during Cretaceous and Tertiary times was partially eroded and deposited in a series of lakes and deltas along the foot of the Piedmont area. They are believed to have been originally quite impure, but had iron and other cations removed by leaching processes during the Tertiary period. The Cretaceous clays are very pure, and thus used mainly in the paper-coating industry; but some of the less pure Tertiary clays are available for decorative tile production.

5.9 EXPANDED CLAY

The process which results in the disastrous bloating in brickmaking (see above) can be turned to advantage in the production of **expanded clay**, which can be incorporated into concrete as a lightweight aggregate. In this process pellets of clay are fed into a furnace at about 1200°C so that rapid vitrification of the outer layer of the pellet occurs before the gases generated in the interior are able to escape. The pellet is thus inflated to many times its original size; and the spongy inner texture results in a low specific gravity – densities as low as 0.3 can be achieved.

It appears that the evolution of sulphur oxides from pyrite, and of carbon dioxide from carbonates, is less important in this process, since these reactions take place at relatively low temperatures; the important reaction is the reduction of iron compounds in the presence of carbon, producing carbon dioxide and oxygen.

Rapid vitrification is important, so the presence of some 20–30% of fluxes in the clay is essential. Organic carbon in excess of 3%, and pyrite in excess of 2%, inhibit the process.

While inherently weak as an aggregate, its light weight allows it to be formed with cement into large blocks which find extensive use in interior walls and floors.

Perhaps the largest production of expanded clay for this purpose is in Belgium, where the Tertiary **Boom clay** is used. This is of Middle Oligocene (Rupelian) age, a stage poorly developed, or represented by sands in Britain and the Paris basin, but which has a marine facies extending from northern Belgium through north-west Germany into Denmark. The clay content of this formation consists almost entirely of smectite and illite, with only minimal amounts of kaolinite, so that its fusion temperature is quite low. The clay also contains up to 4.5% organic carbon, up to 4.3% calcite, and up to 2.5% pyrite. The gases evolved at comparatively low temperatures from these minerals provide the mechanism for the inflation of the pellets; since the atmosphere inside the pellets is entirely a reducing one, such pellets are always black in colour. Sometimes iron-rich metallurgical wastes are added to facilitate the iron/carbon reduction.

Although experimental work has shown that many British clays currently used in brickmaking could be used for this purpose, the dominant occurrence of kaolinite in many of them produces high vitrification temperatures which lead to high energy costs. In Britain and in the United States much of the expanded clay produced is made from coal-measure shales derived from old colliery spoil heaps, sometimes combined with pulverized fuel ash – the cheapness and availability of the raw material thus offsetting the high energy costs. There is, however, one manufacturer who uses Teritary London Clay – which has some similarities to the Belgian Boom clay – as his raw material base.

Here, and elsewhere in the world, lightweight concrete is more commonly produced by foaming, or by incorporating other lightweight fillers such as perlite (p. 192 below).

Another lightweight particle, distantly related to expanded clay, is the **coenosphere**. In coal-fired electricity generating stations, the flues trap much fine dust, which, under the name of pulverized fuel ash or fly-ash (pfa) finds extensive use in the construction industry. Within it there occur a proportion of hollow glassy spheres, which probably have their origin as bubbles of fused clay from the argillaceous matter in the coal. Separated by floating, they can be incorporated into concrete as a substitute for the fine aggregate, producing a lightweight product.

5.10 CONCLUSIONS

Common clay, as used in the manufacture of structural clay products, is thus revealed as a very complex substance, whose mineralogical characteristics have a profound effect on both the process and the product. An understanding of the sedimentology of the deposit is essential if it is to be properly exploited; an understanding of the relationships of raw material to the final product is necessary if that material is to be properly used. A geological approach, which relates all properties back to the raw material, and ultimately to the mode of origin of the clay, has much to contribute to this understanding, and to this large and important industry.

6

Cement and concrete

6.1 HISTORY

British Patent No. 5022, registered by one Joseph Aspdin, of Leeds, in 1824, for 'a superior cement resembling Portland Stone' is surely one of the landmarks in the construction industry; for it introduced a material, which, when combined with aggregate, makes high-strength concrete. It is difficult to visualize what our present-day urban landscape would be like without concrete; or to imagine how today's major engineering structures – bridges, tunnels, roads, high-rise buildings – could have been constructed without it.

Aspdin's invention was to grind limestone and clay together into a slurry, and then fire the slurry in a kiln. The basic idea was not entirely new, since it had been known from 1796 that burning of argillaceous limestone nodules produced a cement which would set under water – this was curiously mis-named Roman cement, and its use was widespread around 1800 – in England it was made by burning the septarian nodules washed out of the London clay along the Essex and Kent coasts. At the same time hydraulic cement was being made by burning argillaceous limestones such as the lower Liassic Blue Lias in England, and similar Jurassic and Cretaceous strata on the continent of Europe.

In fact the Romans had, around 100 BC, discovered a different ingredient – the volcanic ashes of the village of Pozzuoli, on the slopes of Mount Vesuvius, which when mixed with slaked lime and wetted, produced a hydraulic cement – hence the term pozzolan used to describe any material which reacts with lime to form a cement. In this they were preceded by the Greeks, who used volcanic tuffs from Santorini. The use of volcanic ashes in this way continued in all parts of Europe, and the Pleistocene volcanics of the Rhine rift – called trass in Germany and tarras in Holland – were widely exported. They were used, for instance, in 1756, for the construction of the second Eddystone lighthouse. However, such materials must have been expensive, and in restricted supply, so that Aspdin's discovery, using two very common, indigenous materials, was indeed momentous.

6.2 RAW MATERIALS FOR PORTLAND CEMENT

The two basic materials, limestone and clay, in fact often occur in juxtaposition. Between 18% and 25% of clay is usually required, less if the limestone itself has an argillaceous content.

In Britain the limestones used are of Carboniferous, Jurassic and Cretaceous age (Figure 6.1). Where the thick, often chemically pure, Carboniferous limestone occurs, and is immediately overlain by the basal shales of the Upper Carboniferous, the two are exploited in adjacent quarries, and the large works at Clitheroe, Lancashire, Hope in Derbyshire,and Caldon in Staffordshire were sited for this reason on the junction of the two formations. At Dunbar in Scotland, the Carboniferous sequence is one in which there is an alternation of limestone and shale in the same quarry, so that the two can be exploited in the same quarry.

In the Jurassic, the so-called hydraulic limestones, at the top of the Lower Lias, are extensively exploited. The deposit consists of an alternation of limestones and shales, so that by careful quarry design the correct proportions can be maintained. However, works in Lincolnshire and Nottinghamshire are now closed, while those at Rugby use imported chalk; only the South Glamorgan works continue to use this formation.

The Cretaceous chalk is, however, the provider of the largest volume of limestone for the cement industry in Britain. It has the advantage of being uniform, very thick, and soft enough to quarry very economically. The oldest works, on the Thames and Medway estuaries, used a mixture of chalk and estuarine mud, dredged from barges from the mudbanks of the adjacent rivers. These muds are, however, of very variable composition, and the modern successors of these early works use either London clay from Essex, or Gault clay from Sussex. This latter also provides the clay for chalk-based works adjacent to the South Downs in Sussex. Another centre of production is Humberside, again on the outcrop of the chalk, here resting on the Upper Jurassic Ampthill clay, which provides the argillaceous content. Similarly another works sited on the chalk outcrop in Wiltshire uses the immediately underlying Kimmeridge clay.

Locally, however, the argillaceous content of the lowest part of the chalk, the chalk marl and lower chalk, is sufficiently high for it to

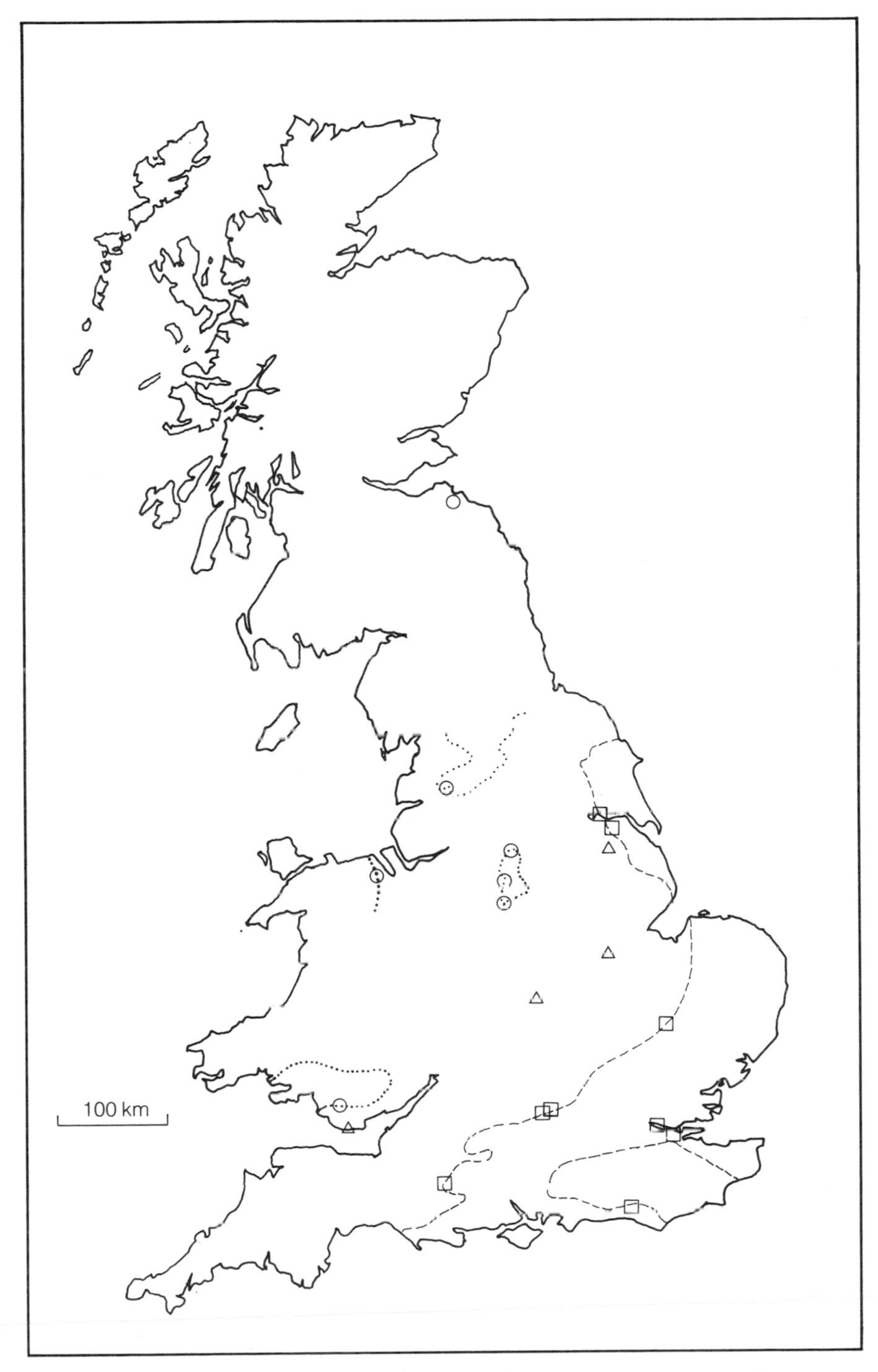

Figure 6.1 Location of cement works in England and Wales in relation to major limestone/shale junctions. □ Chalk; △ Jurassic; ○ Carboniferous; . . . . . . base of chalk; ---- top of Carboniferous limestone.

form the sole source for several works situated at the base of the chalk escarpment in Buckinghamshire, and in Cambridgeshire.

There are very few countries in the world where some limestone and clay cannot be found in sufficient quantities to make cement, and a survey of the world industry would show sedimentary formations from Pre-Cambrian marbles to Recent coral sands forming the basis. Where limestones are locally scarce, ultrabasic igneous rocks (as in Greece) and carbonatite intrusions (as in Malawi) can be utilized.

6.3 MINERALOGY OF THE RAW MATERIALS

Chemical purity does not seem to be an essential for the limestone element of the raw material, although in general limestones which are quite pure are used. It seems that some, but not much, magnesia can be tolerated, but dolomitic limestones are not utilized. Most of the clays added to the mix are marine in origin, and in addition to providing the alumina and silica required to make the cementitious minerals, they also usually contain iron and sulphur – mostly as marcasite or pyrite. Some iron is necessary, providing a flux which lowers the temperature of fusion in the kiln, and which enters into at least one of the important cement minerals, tetracalcium aluminoferrite. Some of the sulphur is oxidized in the kiln, producing sulphur trioxide, which immediately combines with the abundant calcite to produce gypsum; but an excessive amount of sulphur cannot be tolerated, and is a limiting factor in some of the clays which are used.

6.4 CEMENT PROCESSING

Originally the process was 'wet', i.e. a fluid slurry of ground limestone and clay was mixed with water and then fed into the kiln; this water had to be evaporated before the mineralogical reactions could begin. Rising energy costs have resulted in this process being replaced by a 'dry' feed, which seems to achieve the same result with substantially less fuel. The mixture is raised to above 1280°C, at which the minerals are fused to a glassy state, and allowed to cool slowly, so that crystals can develop. The resulting minerals are many and complex, but the chief cementitious ones are **tricalcium aluminate** (abbreviated to C3A); **tricalcium silicate** (C3S); **dicalcium silicate** (C2S); and **tetracalcium aluminoferrite** (C4AF). There is also frequently a percentage of calcium oxide. The resulting clinker is ground to produce cement; at this stage up to 5% ground gypsum or anhydrite is added – this acts as a setting retardant when the cement is hydrated.

At this stage other cementitious materials may be added, giving different products for a variety of different uses. Such materials may be natural – such as volcanic ashes and tuffs, chert and diatomite, which provide additional silica; or artificial compounds such as blast-furnace slag and pulverized fuel-ash (fly-ash or pfa) from power stations – these provide both silica and alumina.

6.5 CONCRETE PRODUCTION

Concrete consists of an inert mineral filler (aggregate) bound together by a binder (cement paste); water is added to cause the hydration of the cementitious minerals, and to allow the concrete mix to be poured or shaped. Usual proportions of these ingredients are water 14–22%, cement 6–18%: the aggregate thus makes up 60–80% of the mix – of this about one-third is fine aggregate and two-thirds coarse.

There are many variables in the formulation of a satisfactory concrete, the precise mineralogy of the cement, the petrography, size distribution, and shape of the aggregate; and the composition of the process water. Further complications are created by the addition of other chemicals as an aid to the flow of concrete (plasticizers), or as an aid to setting. With so many variables it is difficult to predict precisely the properties of the final concrete, and if the quantities to be made are large – for example, in the feed for concrete products, or for a major civil engineering project, it is normal to prepare test specimens of a variety of mixes and to ascertain the crushing strength etc. of these.

Of the variables, the cement constitution is most readily controlled; in the first place it is usually made from raw materials – thick, pure limestones, and uniform marine clays – which do not vary greatly; secondly, routine chemical analysis of the ground clinker can ensure consistency of the product from the works. The uniformity of the aggregate supply is less easy to control, partly because the rocks in the quarry tend to be more variable, partly because of the inherent problems of proper sampling of such bulky material, and partly because of the complexity of the tests needed to recognize the variation. A good example might be that of a dolerite or basalt quarry – depth of weathering in such rocks is notoriously variable, and the product of a quarry may yield a varying proportion of weathered and unweathered aggregate. Only by careful geological supervision of the quarry, and regular petrographic examination of the product, can this be avoided.

It is recognized that concrete requires a length of time to acquire its full strength – a curing period – during which the hydration process is taken to completion, and the excess water evaporated. Concrete products such as pipes and tiles are cured at controlled temperatures and humidities, so as to ensure the development of maximum strength; but the concrete of buildings and structures must be exposed to the vagaries of the local climate.

At this stage shrinkage takes place, and even apparently sound concrete becomes penetrated by a network of microcracks. These microcracks usually develop in the cement paste, or along the cement/aggregate interfaces; but it has been suggested that if crushed rock aggregate is used, microcracking can also extend into the planes of weakness of the aggregate.

Excessive shrinkage, leading to extensive cracking, and possibly eventual disintegration, is associated with certain types of aggregate – these are ones which contain flaky and platy minerals such as mica and clay – these appear to absorb excessive water on mixing, which they then lose on drying, with consequent stress developments within the concrete. This shrinkage can be effectively measured on laboratory samples. Aggregates of quartz, flint and marble can produce shrinkages as low as 0.025%, while aggregates with substantial amounts of shale, mudstone and greywacke may result in values in excess of 0.085% – this will cause rapid deterioration of the concrete in work. In between these extremes varying degrees of shrinkage are related to the presence of greywackes, phyllites, mica-schists and weathered basic igneous rocks.

Concrete has been regarded as an artificial rock, with all the qualities of natural rock, for which it has, in the last century at least, become a major substitute. But there are important differences. First, the minerals in the cement paste are hydrated silicates which do not occur in nature, and therefore whose long-term stability is unknown. Secondly, the necessary drying process creates a rock mass permeated by flaws, and thus vulnerable to weathering processes; and thirdly the hydration processes produce alkaline fluids within the concrete which create an environment for chemical change.

Most natural rocks have been consolidated, by deep burial within the earth's crust, under very considerable pressures, and have thus developed a strength which can only be simulated in concrete by vibrating, and above all by incorporating steel reinforcement. Concrete is high in compressive strength, but weak in tension, and any concrete structure needing to resist stress is provided with abundant steel reinforcement, as rods, meshes and cables.

The residual fluids within the concrete mass after hydration and drying have been completed are strongly alkaline, and this serves to protect the steel reinforcement from electrolytic effects, which would initiate rusting. However, because of the system of microcracks, this situation can be destroyed by the entry of acids. These originate in a variety of ways. Atmospheric sulphur dioxide and nitrogen oxides are one cause; they are of course particularly concentrated in polluted atmospheres, and are much enhanced by power station emissions and by car exhaust fumes – so urban motorways are much endangered by these conditions. Another cause is the liberal use of salt for de-icing roads, which introduces the chloride ion to the steel–cement interface; salt of course can also enter concrete structures from airborne spray, or from saline groundwater – this latter having been a particularly troublesome element in concrete structures in the Middle East. The chloride ion has also in the past been introduced deliberately, as calcium chloride was extensively used to promote quick setting until it was prohibited (in Britain in 1977). All these causes promote rusting of the steel, and the consequent expansion causes large cracks to develop in the concrete along the lines of reinforcement.

Concrete, of course, is subject to the same natural weathering processes as stone (section 2.1) when used in an outside situation. Thus the effects of temperature change, of freeze-and-thaw, and of salt and gypsum intrusion are often manifest. From this point of view, concrete should be regarded as poorly durable, given the difference of expansion properties between the aggregate and the cement paste, and the prevalence of microcracking.

During the last few decades an increasing number of concrete failures have occurred which have proved to be the consequence of reactions between the cement paste and the aggregate particles. Since these have effectively destroyed some major structures – dams, motorway flyovers, car parks, and hospitals – they have received much publicity. Some degree of aggregate/cement interaction appears tolerable, or even favorable, as providing a weld which increases the strength of the concrete. But most of these interactions involve local expansion, and are thus very harmful. They are all concerned with the presence of alkali in the residual fluids; these are formed during the setting reaction of the cement, which releases sodium, potassium, and hydroxyl ions into aqueous solution.

Three distinct processes are known to be involved; they are the

alkali–silica reaction, the **alkali–silicate reaction**, and the **alkali–carbonate reaction**.

(a) Alkali–silica reaction

This is a reaction between the aqueous fluids and the silica of the aggregate particles (although some silica may also be provided by the paste). It results in the formation of a silicate gel which is highly expansive, exerting forces which have been estimated as being as high as 14 MPa – clearly enough to disrupt the strongest concrete. The reaction occurs when the silica is in the form of microcrystalline tridymite and cristobalite or is amorphous; fully crystalline quartz is quite unaffected. The expansion is greatest and most rapid with opal, but it is seen to a lesser degree in silicified carbonate rock, in cherts and in greywackes etc. (Figure 6.2). It has also been observed where cryptocrystalline quartz is an accessory mineral in some andesites and granites.

Silica reactivity can be tested by making up small discs of aggregate embedded in cement which are then immersed in a strongly alkaline solution; the silica gel formed by the reaction can be seen on the surface after a few days.

(b) Alkali–silicate reaction

The aggregate involved in this reaction is that containing platy or layer-structure minerals, such as phyllites, greywackes, chlorite schists and the like. The mechanism is not so well understood, but it is believed to be related to the expansion of the silicate layers on contact with the alkaline liquid.

(c) Alkali–carbonate reaction

Aggregate consisting of non-dolomitic limestone, and pure dolomitic rock, reacts with the cement paste to form a reaction rim to the aggregate particle which is seemingly beneficial. If some clay mineral is present in a dolomite a different reaction occurs; the margins of the dolomite fragment become dedolomitized with the formation of the mineral **brucite** ($MgOH_2$), which then reacts with the silicate ions. This dedolomitization creates microcracking, which allows water to penetrate to the clay minerals, whose swelling creates an overall expansion, which thus disrupts the concrete.

In view of the scale of the damage caused by these reactions, it is not surprising that there is an intense search for remedies and

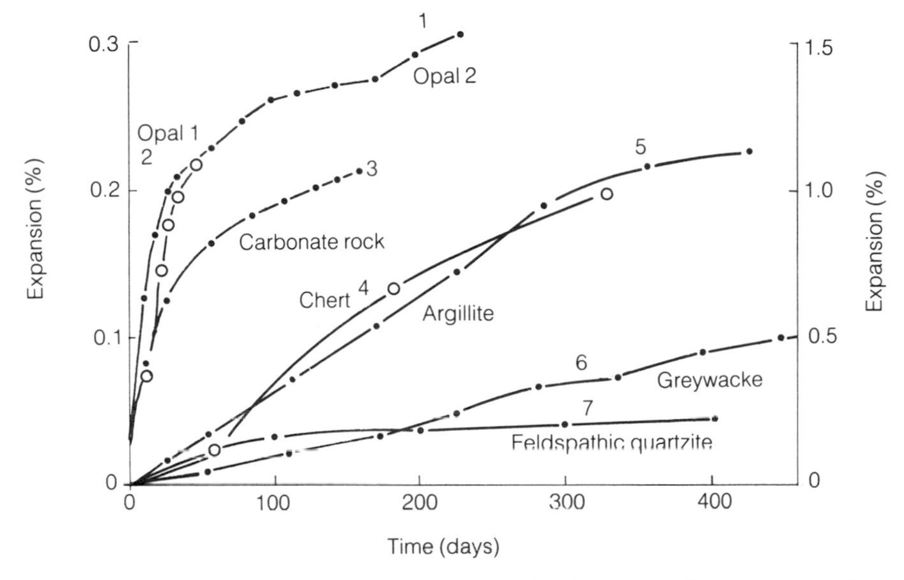

Figure 6.2 Expansion (%) against time (days) curves for some reactive aggregates (after French, 1980). The expansion scale for Vivian's experiments on opal is five times that for the other aggregates. 1, opal (Vivian, 1950); 2, opal (Hobbs, 1978); 3, Hadlitha carbonate aggregate (Alsinawa and Murad, 1976); 4, chert (Ozol, 1975); 5, argillite (Dolar-Mantuani, 1969); 6, greywacke (Dolar-Mantuani, 1969); 7, feldspathic quartzite (Gratton-Bellow and Litvan, 1976).

preventatives. Measures so far undertaken are (a) attempts to exclude water from the concrete, by the use of waterproof coatings and claddings; and (b) the addition of various chemicals and minerals (e.g. silica fume, zeolites) to the cement to reduce the effects of the alkaline liquids. Until such measures are shown to be wholly successful, however, careful selection of aggregate to avoid reactive materials must be the major consideration. This has, however, the effect of severely restricting the availability of aggregate supplies; and there may be many parts of the world where the supply of a totally non-reactive aggregate will prove to be quite impossible.

6.6 CONCRETE BLOCKS AND RECONSTITUTED STONE

In contrast to the sophistication of modern concrete technology, the concrete block is the simplest possible building unit. Sand, cement and water mixed together, poured into a mould and left to set, then demoulded and dried – such is the simple process which provides a building unit for millions of people throughout the developing world. Such blocks have poor durability, poor insulation from heat and noise, and low strength – but they are cheap, and the major component – sand – is likely to be available in the locality.

Insulation can be improved by casting the unit as a hollow block, and by the use of lightweight and other aggregates in the mix. Thus the factory-produced concrete block of the advanced countries may be a very different kind of product. The same considerations on selection of aggregate of course apply as with mass concrete if such blocks are to have a high frost resistance and long life.

The high cost of natural stone in recent years, and the need for new houses to blend with older, stone-built ones, has led to the development of reconstituted stone. To make this, crushed rock is incorporated into a cement mix, cured and dried. Often the block is made to double size, and split down the middle to give the appearance of naturally broken rock. By using local stone for the base – for example, oölitic limestone in the Cotswolds, Millstone Grit in the Pennines – a passable imitation of natural rock can be achieved; although in one instance a manufacturer found that use of an oölite-rich alluvial sand gave a better product than the crushed oölite. What is more difficult to achieve in a machine-made product is the random sizing of natural stone, and a too-regular pattern of blocks and joints immediately reveals the origin of the material.

6.7 BASE-COURSE STABILIZATION

In areas of tropical weathering, aggregate of sufficient strength to provide a base-course for roads may be hard to find. In these situations, stabilization of the local soil with lime or cement can provide a satisfactory substitute. It has been shown that soils which have a wide range of particle size require the least cement (as little as 3.5% in some cases); while materials with a limited size range, such as medium sand and clay, require the most (up to 12%). Higher additions of cement give increased strength, but unconfined compressive strengths of 6000 $kN\,m^{-2}$ are achievable. The fundamentals of the process are not really understood, but it seems likely that something more than simple void-filling is involved.

6.8 CONCLUSIONS

Cement, more particularly modern Portland cement, is a highly versatile material with uses ranging from the simplest to the most technologically advanced. The urban civilisation of today could hardly exist without concrete; and the discovery that its durability can be severely limited has come as a shock and surprise to architects and engineers. It should not, however, be a surprise to geologists. Concrete is an artificial compound of largely reactive minerals; compared with natural rock it could not be expected to be durable. Natural rock has been used for several thousand years for construction, and naturally exposed for millennia, so that we have had a good opportunity to study its behaviour. With hardly more than a century's observation possible on concrete, it is not surprising that we have much to learn.

7

Minor construction materials

The variety of materials used by the construction industry is enormous – and almost all these materials are, in the first instance, based on minerals. All metals and most of the ingredients of plastics are manufactured from minerals. In many of these, however, the connection between the geological occurrence and the finished product is tenuous. In the sections which follow, some of those materials are described where geological factors have a direct bearing on their manufacture and use.

7.1 GLASS

Glass was hardly an important part of a building until the late eighteenth century, when improvements in casting techniques allowed the production of fair-sized sheets suitable for windows. The Crystal Palace, built for the Great Exhibition in London in 1851, was the first building in which glass was a major element. Very large sheets of glass are now produced by rolling or drawing, or by floating on molten metal. The other constructional use is as glass fibre, which is used on its own for insulation, and incorporated into a resin body as fibreglass.

The chemical composition of window glass lies within the following limits: SiO_2, 70–74%; Na_2O, 12–16%; CaO, 5–11%; MgO, 1–3%; Al_2O_3, 1–3%. The essential element is silica, which can be fused to glass on its own, although a temperature in excess of 1700°C is required. Sodium is introduced to lower the fusion temperature; with the above proportions a melting temperature of around 1200°C is required. Higher amounts of soda addition produce the water-soluble water glass. Without the calcium, magnesium and alumina the glass can be unstable and weak.

The chief impurity to be avoided is **iron**. Minute amounts of this will colour the glass green, larger amounts will turn it brown. For clear flat glass the maximum tolerated is 0.18% Fe_2O_3, and less than 0.1% is preferred. For glass fibre, Fe_2O_3 contents marginally higher than 0.3% are acceptable. Since iron minerals are almost ubiquitous in nature, this requirement severely restricts the mineral deposits which can be used for making glass.

The sand used for glass-making is usually natural aggregate, although rarely crushed rock is used. There are close specifications for grain-size distribution, for mineralogy and for chemical purity.

The grain-size distribution is critical in the melting process, and needs careful control to ensure even fusion and to avoid 'stones' (i.e. areas of poorly melted glass) and bubbles. Specifications have tended to become tighter as the demands of glass technology increase (Figure 7.1); they are very precise is excluding both coarse (not more than 0.25% retained on a 710 μm sieve) and fine (nothing passing a 90 μm sieve). Grading of the middle range of sizes is not defined, but needs to be even. Comparison of Figure 7.1 with the size characteristics of natural sands (Figure 4.7) shows that few natural sands have gradings which lie exactly within these limits, so that sand must be washed (usually by hydrosizer) to provide this grading. A further restraint on the undersize is the requirement that the alumina content must lie below 1.4% – since the alumina in most non-felspathic sands is in the clay fraction.

The ideal glass-sand would consist of 100% quartz grains but such sands do not occur in nature. The closely defined alumina limits exclude any felspathic sands. The low iron content demanded means that all iron-containing minerals must be eliminated. A further group of undesirable minerals are the **refractory minerals**; of these chromite and chrome spinel are regarded as the most objectionable; but tourmaline, staurolite, monazite, zircon, etc. – all the common heavy minerals of sedimentary rocks – all have higher melting points than silica, and thus create uneven melting in the glass. The size of these particles is very important; whereas the industry will tolerate 1640 particles per tonne at the 'passing 212 μm' level, it will only allow 20 particles per tonne if they exceed 250 μm, and none at all in excess of 355 μm. These figures are extremely low, representing only 0.000 0004% of the sand, and are a severe constraint on the producing industry.

The occurrence of a totally iron-free sand requires very special geological conditions (see below), so that most sands need to be processed. Washing and hydrosizing will extract some iron, if it is finely divided. Frequently the sand grains have adherent clay films, and iron-mineral coatings, and this can be removed by **attrition scrubbing**. More strongly adherent coatings, and some iron-

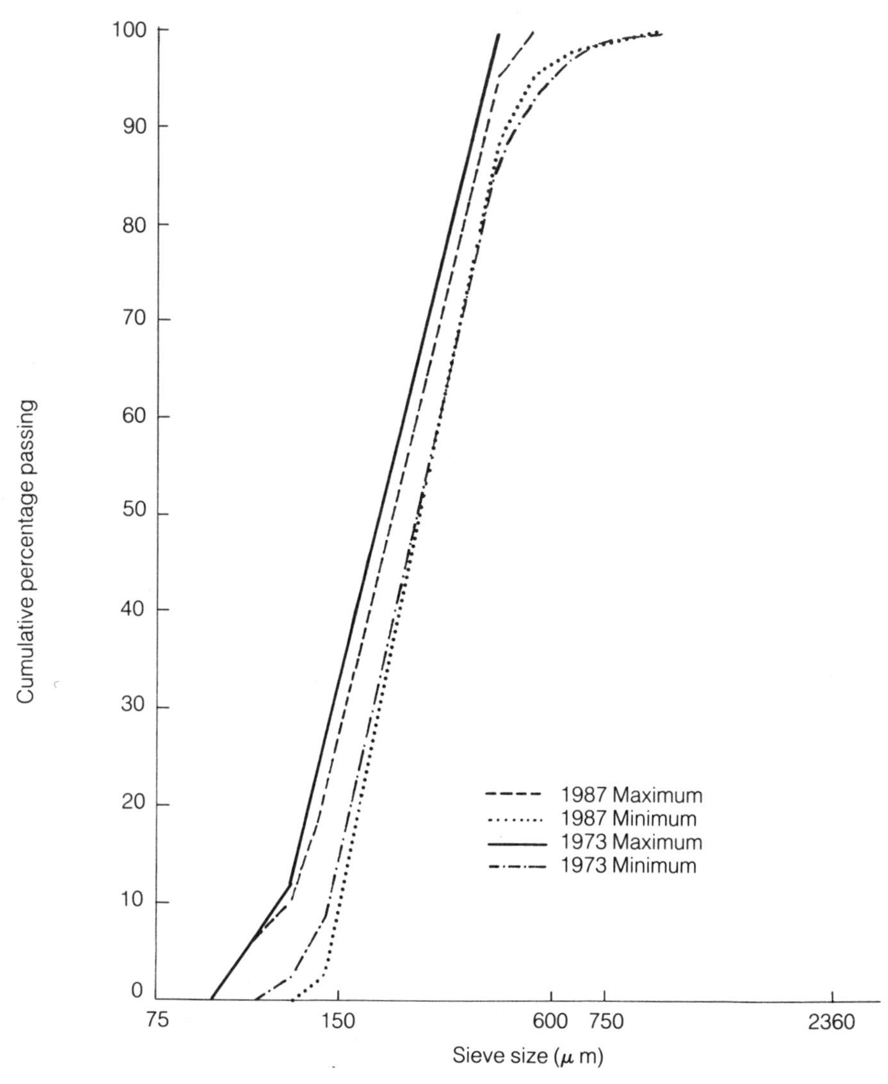

Figure 7.1 Grain-size distribution limits for glass-sand in 1973 and 1987 (after Kirby and Lavender (1987).

mineral grains, can be removed by **acid leaching**, using either hot or cold hydrochloric acid. Finally, **magnetic separation**, either wet or dry, can be used to remove iron mineral grains. Costs of this processing are high, and the removal of large percentages of iron both difficult and uneconomic; so that sources of glass-sand are still mostly sought where iron levels are naturally low.

With the exception of some very small exploitation of vein quartz, all glass-making sands are of sedimentary origin. Iron minerals are so common that circumstances in which iron-free sand are formed are extremely unusual. It is useful to distinguish two categories: those sands which were deposited in an iron-free form, and those which contained iron which has subsequently been removed.

The depositional environment for the first category must be one in which the source area is granitic or gneissose (to supply the quartz) and which is subjected to chemical weathering (to reduce the silicates to clays); the depositional environment has to be such that the iron-rich clays are then removed by very efficient sorting. These conditions were locally realized in the west of Scotland during the Upper Cretaceous, forming the Loch Aline sandstone; this, and a similar deposit at Provodín in Czechoslovakia, were marine coastal sands adjacent to an area from which virtually no argillaceous sediment was being derived. It is likely that the Tertiary white sand formation of Guyana (which represents one of the largest unexploited reserves of glass-sand in the world) was formed in a similar way.

Many more glass-sands are deposits formerly containing iron; the principal agent in its removal has been downwardly percolating water, particularly if that water was acidified by growth and decay of vegetation. Thus in the Carboniferous, where coal seams have formed on top of sandstones, the sandstones have been leached of their iron; these ganisters have been important sources of silica sand for glass. In the Cretaceous of the north of the Paris basin, thick fine-grained sandstones and sands contain interbedded seams of lignite and lignitic clay; groundwater percolating downwards from these lignites has removed the iron to lower levels. Geologically more recent examples come from the Pleistocene blown sands of south Lancashire (the Shirdley Hill sand which forms the original basis for the concentration of glass-making in this region), Lincolnshire, and northern Belgium. Extensive peatbogs formed on top of these sands in late Pleistocene times, and where they do so the sands beneath are strongly leached. This leaching has sometimes, as in Cheshire, extended into the older Triassic sandstones beneath. In the south of England, the Lower Cretaceous greensand deposits originally contained much iron in the form of glauconite; but the top layers have been leached of this, and the iron removed to a thick ironstone layer at depth. This is clearly related to a higher groundwater level than at present, suggesting that it is of late Pleistocene age; and the efficiency of leaching is probably related to the acid waters percolating from the highly pyritic Gault clay, which immediately overlies the sand.

It follows that, as conditions for glass-sand formations are often local and special, variations within such deposits occur; this particularly applies where the low iron content is due to local leaching. It is common, therefore, for sandpits to devote only part of their production to glass-sand, selling the inferior grades as foundry or building sand.

The sodium content of glass is today provided almost entirely by the addition of sodium carbonate, which may be natural or artificial. It occurs naturally as the mineral **trona** ($Na_2CO_3 \cdot NaHCO_3 \cdot 2H_2O$), which requires only calcination to convert it to anhydrous **soda-ash**. Commercial deposits occur in Wyoming and California in the United States, and in the African Rift Valley of Kenya; they are evaporite deposits of Tertiary age. Normally evaporite deposits contain sodium as the chloride halite, and special conditions are necessary for trona to be formed. These conditions include intense volcanic activity (to provide the sodium and carbon dioxide), and much decaying vegetation. Sodium carbonate can also be extracted from brines, as in Mexico.

Apart from the United States, most of the soda-ash used in industry is produced by the **Solvay process**, in which rock-salt and limestone are calcined together with an ammonia catalyst to produce soda-ash and calcium chloride. Whether natural or artificial, soda-ash is produced as a pure chemical product whose addition to the glass mixture can be readily controlled.

The other elements needed in constructional glass are aluminium, calcium and magnesium. A variety of other minerals are added to other glasses for other purposes – lead for crystal glass, boron for heat-resistant glass, etc.

Most glass-sands contain some alumina, and manufacturers will accept up to 1.4% in the sand. It is most important that the alumina content should not vary, however, and there is an additional specification that variation from the mean should not exceed ±0.15%. Since up to 3% of alumina is required in the product, this must be added, taking care not to introduce further iron at the same time. Favoured minerals, which can be obtained in a very pure form, are kaolinite, feldspar, and nepheline syenite.

Calcium can be introduced by limestone. The conditions under which limestone was formed were often those of clear tropical seas, in which the only sediment was provided by the skeletons of organisms, or by chemical precipitation. So chemically pure limestones are not uncommon at all levels in the stratigraphical column. In Britain, the Carboniferous limestone, and the chalk, are major providers of pure limestones suitable for glass-making.

Finding a suitable dolomite is less easy. Dolomite, which is a double carbonate of magnesium and calcium, can be used to provide both these elements if an iron-free deposit can be found. Unfortunately, dolomite forms a solid solution series with iron compounds to produce minerals such as **ankerite**, which commonly occur along with it, and which cannot readily be separated. Pure dolomites are rare; in Britain only very selected parts of the extensive Permian magnesian limestone can be used, and most of the dolomite used is imported from Spain. The industry also uses **magnesite** as a source of magnesium; this can be found associated with ultrabasic vulcanicity, for instance in Greece, or can be manufactured from dolomite and sea-water.

7.2 GYPSUM

Calcium sulphate dihydrate ($CaSO_4 \cdot 2H_2O$) occurs in nature in three forms: as the clear crystalline form **selenite**; as the massive form **alabaster**; and as the fibrous, massive form **gypsum**. The anhydrous form **anhydrite** ($CaSO_4$) occurs in association with gypsum.

Alabaster has had a long history of use as an ornamental stone; it is much softer than marble, and therefore it is easier to carve; but its softness makes it unsuitable for use as floor tiles or any place where there is substantial wear. It is also difficult to obtain large blocks without flaws, thin veins and fragments of marl being frequent. It was formerly mined from the Upper Triassic of Derbyshire; but today the only commercial source is from the Miocene strata of Tuscany. It occurs as a local facies of gypsum deposits, and is found in small quantities wherever they occur.

Gypsum is the commonest form of occurrence, and the only one of significance for construction purposes. When heated to 107°C it loses three-quarters of its water, becoming the hemihydrate ($CaSO_4 \cdot \frac{1}{2}H_2O$) – plaster of paris. It can then be mixed with water, and either by itself, or in combination with fine aggregate, used for rendering walls and ceilings. The speed of the hydration is very rapid, so usually a variety of organic chemicals are added to retard the setting time. In recent years the largest consumption of hemihydrate has been in the production of wallboards, in which hydrated plaster is compressed between two layers of heavy paper; this is mainly used for lining internal surfaces, providing a smooth and rigid base upon which a rendering can be applied.

Gypsum occurs generally as veins and seams in argillaceous deposits, from which it is extracted by selective quarrying. It has generally a pure white colour, so is readily separable from the dark matrix; the only impurities are argillaceous fragments. It is more difficult when anhydrite is present, as this is almost impossible to distinguish visually from crystalline gypsum. It does not, of course, dehydrate, and does not take part in the rehydration process, so that its presence can reduce the quality of the plaster.

In fact quality considerations do not appear to be very exacting. The only British Standard Specification requires:

1. a sulphur trioxide content not less than 35% (pure hemihydrate has 56%)
2. a calcium oxide content not less than two-thirds sulphur trioxide content (for pure hemihydrate it is 70%), and
3. loss on ignition between 4% and 9% (for pure hemihydrate it should be 4.4%).

The last requirement would, however, effectively exclude plaster which contained anhydrite in any quantity.

Gypsum is invariably of sedimentary origin, occurring as one of the many evaporite minerals which originate either by the evaporation of saline lakes, or by saline/groundwater interactions at their margins. In consequence it occurs interbedded with shallow water sediments, marls and sands, and with other evaporites. It was probably deposited originally as anhydrate, but this rapidly became converted to gypsum – the associated volume change often producing complex structures in the adjacent sediments. It is, however, moderately soluble in water, and so is only preserved at or near the surface where it has not been exposed to extensive groundwater percolation – i.e. where it outcrops in regions of low rainfall, or where it has been protected by inclusion within an impermeable sequence.

Gypsum deposits occur throughout the world at a wide range of geological horizons: Cambrian in Pakistan, Silurian in New York and Ohio, Devonian and Mississippian in the mid-western United States and British Columbia. But in both America and Europe, the largest accumulations of gypsum were in the Permo–Triassic (Figure 7.2).

During the Carboniferous and Permian north-central Europe was occupied by a shallow sea, which was subject to periodic desiccation, with consequent increase in salinity. It is now recognized that there were many periods of hypersalinity during the Carboniferous, but the evaporite deposits have been very largely dissolved away, probably by the next marine incursion. They have been found intact in deep boreholes in Belgium, and recently a workable deposit of this age has been found in the north of the Irish Republic.

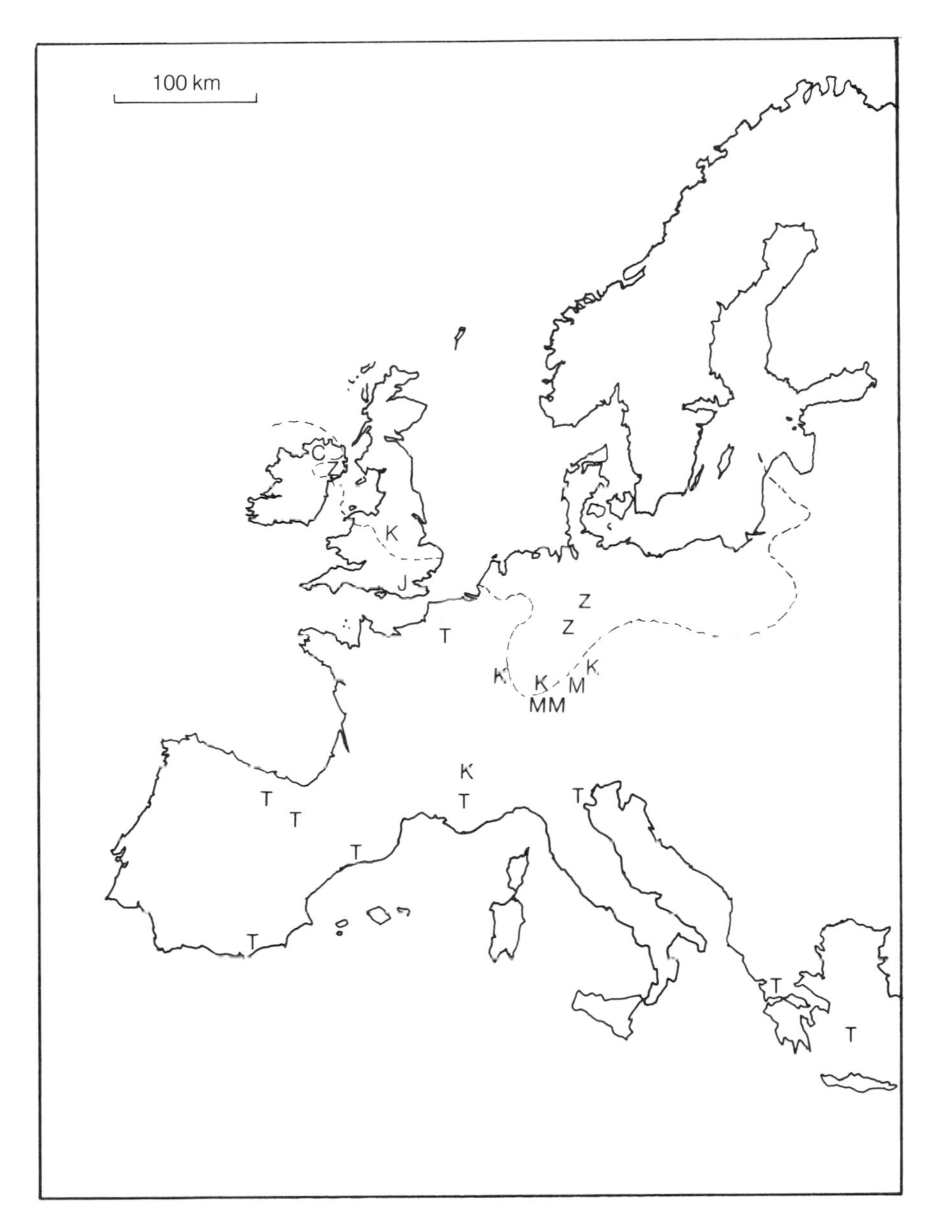

Figure 7.2 Sources of gypsum in western Europe in relation to palaeogeography. T, Tertiary; J, Jurassic; K, Keuper; M, Muschelkalk; C, Carboniferous; Z, Zechstein; ----, southern margin of Zechstein Sea.

Desiccation became more constant in the Permian, and the **Zechstein Sea** had a more restricted extent. Within it thick evaporites accumulated, and they are worked for gypsum in the regions marginal to the Harz mountains. These conditions persisted into the Triassic, with workable gypsum deposits at Middle Triassic (Muschelkalk) and Upper Triassic (Keuper) horizons. It is these latter horizons which are worked in Britain, in Nottinghamshire and Cumbria.

This phase of desiccation was brought to an end by the incursion of the shallow Jurassic sea, and marine conditions prevailed over much of north-western Europe from then on. From time to time local basins became separated, and small evaporite sequences formed. One such was in the latest Jurassic of the south of England, where a workable deposit is found in the centre of the Wealden anticline. Another such provided the many Oligocene gypsum deposits of the Paris basin – the original plaster of paris – still extensively worked today.

By Tertiary times the centre of evaporite deposition was further south; the Mediterranean basin became landlocked during the Miocene, and thick saline accumulations resulted. Many of these contain horizons of gypsum, which are worked in Spain, Italy, France, Greece, Turkey and Egypt. The full extent of these deposits is unknown, but they will clearly be a major source of gypsum for many years.

A new source of supply which is now becoming available is **FGD gypsum**. This abbreviation stands for **flue gas desulphurization**. The somewhat reluctant acceptance by the European Community that acid rain – that is, polluted rain containing sulphuric and nitrous acids – does originate from coal- and oil-burning electricity generating stations, and that it is seriously damaging the environment, has resulted in a general agreement to substantially reduce emissions from these power stations. To do this the flue gases of the power stations must be scrubbed – i.e. the gases must be taken out in aqueous solution. The acidic liquid must then be disposed of – there are currently two possibilities. One is to use the sulphuric acid as a chemical base for sulphur and sulphur compounds. The other is to neutralize the acid with ground limestone, which will produce calcium sulphate – FGD gypsum.

The quantities of gypsum which will be produced are very large indeed, and their impact on the market for natural gypsum must be dramatic. For instance, in West Germany, whose programme is now well advanced, it is estimated that on completion of the scheme, nearly 4 million tonnes of FGD gypsum will be produced annually;

at present total West German consumption of natural gypsum is only about 3 million tonnes. This is not unwelcome in Germany, since the gypsum mining companies have been under great pressures for environmental reasons. In Britain, one major power station alone will produce a million tonnes of gypsum in a year.

Apart from a finer grain size, which causes some problems in handling, the FGD gypsum does not seem to be an inferior product for the traditional uses.

This is not a geological problem in itself; but it has two consequences which are. One is the increased demand for high-purity limestone. The power industry will clearly seek to use limestone of the highest purity, since this will reduce the volume to be handled to a minimum, and improve the purity, and hence the saleability of the product. The sources of such limestones are largely in areas of landscape protection. The second problem is that of disposal of surplus gypsum; if, as seems likely, all the FGD gypsum cannot be sold, it will have to be disposed of to landfill sites. These two problems must involve geological understanding if they are to be properly and effectively resolved.

7.3 INSULATORS AND LIGHTWEIGHT AGGREGATES

There are a number of minerals which can be used to improve noise and heat insulation in buildings; although geologically unrelated, they have in common an open texture which can entrap air, and consequently low specific gravity. They may be used by themselves as a loose filler, as between the rafters of a house roof, or they may be incorporated with cement into a lightweight block.

Pumice is the product of explosive volcanic eruptions. Such eruptions occur where the volcanic lava is siliceous, and therefore viscous, and there is much entrapped water and air. The sudden release of pressure causes a rapid expansion, creating a frothy mass full of bubbles. This cools very quickly in the air, producing a lightweight rock full of unconnected vesicles. Associated with this, the smaller particles form a volcanic ash of glassy fragments, known as **pumicite**. Moulded with cement, and cured slowly, these materials produce a lightweight block with good insulating properties, but relatively low strength.

While pumice may be found in rocks of many different geological ages, those belonging to older periods generally have their vesicles filled with mineral materials, and so do not have the low specific gravity which is the essential value. All economic pumice deposits are located in areas of very recent volcanic activity – that is, the circum-Pacific zone, and the Mediterranean–Himalayan belt.

The pumice of the island of Lipari, off the north coast of Sicily, was worked by the Romans, and continues to be a source of aggregate for the Italian construction industry. The Greek islands, and the Turkish mainland, have extensive deposits. In America, the newer volcanic areas of California, Arizona, Oregon and New Mexico yield large quantities; while large unexploited resources exist in St Lucia and other Caribbean islands.

The explosive products of volcanoes whose rock products are of a more basic composition are known as **scoria**; these are not so light as pumice, and are dark in colour. They are, however, extensively used as road-base materials for minor roads in the western United States, where the very recent vulcanism of the Cascade mountains has provided them in large quantities. When incorporated with cement into a lightweight block, they are reported to be of good insulating quality, and to produce a stronger block – compressive strengths of up to 20 MPa are reported, as compared with a maximum of 10 MPa for blocks made with pumice.

Also of volcanic origin is the material known as **perlite**. Strictly speaking, the term perlite refers to a texture found in volcanic glass, where the glass may be seen to be broken up by numerous curved and spherical cracks. Commercial perlite is the name given to a volcanic glass which can be made into an expanded aggregate. The glass must be rhyolitic – i.e. highly siliceous – and must contain between 2% and 5% water. It is ground to granules and then heated to around 1100°C, at which it expands to up to twenty times its original volume. Like pumice, perlite only occurs in the youngest rocks, since the volcanic glass tends to vitrify with age, and the oldest examples are Oligocene.

Perlites mainly occur on the quenched margins of highly siliceous lava domes. There is an important zone of such recent volcanic activity extending from the Greek islands of Milos and Kos, through Turkey, and into Armenia and Kazakhstan; while in the United States production is mainly from New Mexico.

Whereas pumice and perlite derive from acid volcanic rocks, **vermiculite** is associated with ultramafic rocks – but apparently only when such rocks have been invaded by later silica-rich intrusions. When such complexes are deeply weathered, their ferromagnesian minerals are converted to mica-like minerals. It is probable that more than one mineral species is involved; but all have a sheet-like structure including interlayer water. When heated rapidly to about 880°C, the interlayer water expands explosively, forming a granular, low-density product. Since conditions for its formation are so precise, localities in which vermiculite occur are somewhat limited; but there are very large deposits in Montana in the United States, and in the Transvaal.

Aggregates produced from expanded clay (section 5.9), with pumice, perlite and vermiculite make up the largest volume of lightweight aggregate used by the construction industry today. There are many other variants – natural minerals such as diatomite, and certain zeolites; processed minerals such as expanded micas; and wholly artificial products such as coenospheres and expanded pfa. The advantage of lightness is not only that it makes handling easier, but it reduces the weight of large structures on their foundations. This is combined with better acoustic and thermal insulation. As pressure on supplies of conventional aggregates increases, so industry may well turn more and more to these novel materials. We still have much to learn about the conventional constituents of our buildings; these newer ones present a new challenge to our understanding.

Sources of text-figures

Alsinawi, S. A. and Murad, S. (1976) On the alkali-carbonate reactivity of aggregates from Iraqi quarries. *Symposium on the effects of alkalis on the properties of concrete.* Cement and Concrete Association, London, pp. 255–72.

Barshad, I. (1966) The effect of a variation in precipitation on the nature of clay mineral formation in soils from acid and basic igneous rocks. *Proceedings International Clay Conference 1966*, Jerusalem, Israel, Vol. 1, pp. 167–173.

Brindley, G. W. (1980) Quantitative X-ray mineral analysis of clays in *The crystal structures of clay-minerals and their X-ray identification* (eds G. W. Brindley and G. Brown) Mineralogical Society Monograph 5, pp. 111–138.

Brown, G. J. (1963) Principles and practice of crushing and screening. *Quarry Managers Journal*, March–September 1963.

Decleer, J. (1983) Studie van de relaties tussen chemische, fysische en mineralogische kenmarken van de Boomse klei en van de verhittingsprodukten. Thesis, Kath. Univ., Leuven.

Department of Transport (1976) *Specification for Road and Bridge Works*, London.

Dibb, T. E., Hughes, D. W. and Poole, A. B. (1983) Controls of size and shape of natural armourstone *Quarterly Journal Engineering Geology London*, **16**, 31–42.

Dolar-Mantuani, L. (1969) Alkali-silica reactive rocks in the Canadian shield. *Highway Research Records*, **268**, 99–117.

Edelman, C. H. and Favejee, J. Ch. L. (1940) On the crystal structure of montmorrillonite and halloysite *Zeitschrift für krystallographie*, **102**, 417.

Fookes, P. G. and Poole, A. B. (1981) Some preliminary considerations on the selection and durability on rock and concrete materials for breakwaters and coastal protection works. *Quarterly Journal Engineering Geology London*, **14**, 97–128.

French, W. J. (1980) Reactions between aggregates and cement paste – an interpretation of the pessimum. *Quarterly Journal Engineering Geology London*, **13**, 231–247.

Gratton-Bellow, P. E. and Litvan, G. G. (1976) Testing Canadian aggregates for alkali expansivity. *Symposium on the effects of alkalis on the properties of concrete.* Cement and Concrete Association, London pp. 227–42.

Goodlet, G. A. (1964) The kamiform deposits near Carstairs, Lanarkshire. *Bulletin Geological Survey Great Britain* No. 21 pp. 175–196.

Hartley, A. (1970) The influence of geological factors on the mechanical properties of road-surfacing aggregates (with particular reference to British conditions and practice). *Proceedings 21st Symposium Highway Geology*, University of Kansas.

Hawkes, J. R. and Hosking, J. R. (1972) British arenaceous rocks for skid-resistant road surfacings. *Road Research Laboratory Report* LR488.

Hobbs, D. W. (1978) Expansion of concrete due to alkali-silica reaction: an explanation. *Magazine Concrete Research*, **30**, 215–21.

Kazi, A. and Al-Mansour, Z. R. (1980) Empirical relationship between Los Angeles abrasion and Schmidt hammer strength tests with application to aggregates around Jeddah. *Quarterly Journal Engineering Geology London*, **13**, 45–52.

Kirby, C. and Lavender, M. (1987) Developments in the glass sands supply industry *Industrial Minerals*, **242**, 55–8.

Kreimayer, R. and Eckhardt, F. J. (1987) Relations between the firing colour of bricks and the raw material composition *4th Meeting European Clay Group*. Abstract.

Kühnel, R. A. and van der Gaaste, S. J. (1987) Formation of clay minerals by mechanochemical reactions during grinding of basalt under water. *Proceeding of sixth meeting of European clay groups, Seville, Spain* 1987, pp. 331–333.

McLean, A. C. and Gribble, C. D. (1979) Geology for civil engineers. Geo. Allen and Unwin, London, 310 pp.

National Stone Directory 6th. Ed (1984) Maidenhead, 124 pp.

Ozol, M. A. (1975) The pessimum proportion as a reference point in modulating alkali-silica reaction. Symposium on alkali-aggregate reaction, Reykjavik.

Perrin, R. M. S. (1971) The clay mineralogy of British sediments. *Mineralogical Society (Clay Minerals Group)* London, 247 pp.

Pettijohn, F. J., Potter, P. E. and Siever, R. (1973) Sand and Sandstone. Springer-Verlag, Berlin, 617 pp.

Ramsay, D. M., Dhir, R. K. and Spence, J. M. (1974) The role of rock and clast fabric in the physical performance of crushed-rock aggregate. *Engineering Geology*, **8**, 267–85.

Ridgway, J. M. (1982) Common clay and shale. *Mineral Dossier* No.

22, London, 164 pp.
Rigan, J. & Žabka, A. (1976) Dolomitové kamenivo do betónu *Výskumný Ústav Inžinierskych Stavieb, Bratislava*, **104**.
Road Research Laboratory (1959) Roadstone data presented in tabular form. *Road Note*, 24., London, 8 pp.
Stein, V., Eckhart, F. J., Hilker, E., *et al.* (1980) Untersuchung von Tonen und Tonsteinen auf ihre ziegeleitechnische Eignung. Unveröff Ber. Arch. Nds. LA für Bodenforsch (Nr. 83597) Hannover.
Stein, V. *et al.* (1981) Die ziegeleitechnischen Eigenschaften Niederschsischen Tone und Tonstein. *Geologische Jahrbuch*, **45**, 1–51.
Tucker, M. E. (1981) *Sedimentary Petrology*, Blackwell Scientific, London, 252 pp.
United States Bureau of Mines 1949–56 Annual Reports.
Vivien, H. E. (1950) Studies on cement-aggregate reaction. *Bulletin Commonwealth scientific and industrial research organization, Australia* 256.
Zingg, T. (1935) Beiträge zur Schotteranalyse. *Schweizer Mineralogie und Petrologie Mitteillungen*, **15**, 39–140.

Index

References in italic type refer to Figures.